Congenital Abnormalities of the Optic Nerve and Related Forebrain

Congenital Abnormalities of the Optic Nerve and Related Forebrain

THOMAS E. ACERS, M.D.

Professor and Chairman, Department of Ophthalmology
Director, Dean A. McGee Eye Institute
University of Oklahoma College of Medicine
Oklahoma City, Oklahoma

Lea & Febiger
Philadelphia
1983

Lea & Febiger
600 South Washington Square
Philadelphia, Pennsylvania 19106
U.S.A

Library of Congress Cataloging in Publication Data

Acers, Thomas E.
Congenital abnormalities of the optic nerve and related forebrain.

Bibliography: p.
Includes index.
1. Optic nerve—Abnormalities. 2. Prosencephalon—Abnormalities. I. Title.
RE727.A25 1983 617.7′3 82-24962
ISBN 0-8121-0889-2

PRINTED IN THE UNITED STATES OF AMERICA

Print Number: 3 2 1

This work is dedicated to the memory of Dr. Frank B. Walsh, my mentor: the master in the use of the ophthalmoscope and foremost interpreter of all that it revealed.

Preface

I do not intend this book to be an encyclopedic treatise of congenital abnormalities of the eye and central nervous system. Such monumental works have already been provided us by Walsh and Hoyt, Mann, Duke-Elder, and others.

I would describe this work as a compendium to provide a concise reference to a variety of congenital defects that primarily affect the optic nerve and its embryologically related forebrain. For those who possess the time and inclination to pursue the topics in greater detail, the most relevant and complete references for each subject are listed.

Special emphasis has been placed on Chapter 6, "Optic Nerve Hypoplasia," for several reasons. With the recent technologic developments of echography and computed tomography, we are now able to study the optic nerve in vivo precisely. The syndrome of septo-optic-pituitary dysplasia also demonstrates more clearly the association of optic nerve anomalies with abnormal development of the forebrain.

I express my appreciation to Janet Wolf for her untiring efforts in the manuscript preparation and to Nancy Ryan for her fine illustration work. Dr. Pat Barnes has provided invaluable help in the neuroradiologic studies. Dr. Robert Osher and Dr. J. Donald Gass have provided many of the fundus photographs and critically reviewed the text.

Oklahoma City, Oklahoma THOMAS E. ACERS

Contents

Contents

Section I

The Optic Nerve

Chapter 1

BASIC EMBRYOLOGY AND ANATOMY

EMBRYOLOGY

The short optic stalk connecting the optic vesicle to the forebrain (4-mm stage) is originally open to both cavities, and its walls are composed of a single layer of undifferentiated epithelial cells (Fig. 1–1). As the optic vesicle invaginates (5- to 15-mm stage), the distal stalk shares in the invagination process, and a shallow depression appears along its ventral aspect. The stalk lengthens, thus increasing the distance between the brain and the surface ectoderm; it becomes thinner, and the lumen diminishes in the 15- to 17-mm stage (Fig. 1–2).

During this time, nerve fibers are advancing toward the stalk from the retinal ganglion cells (15-mm stage), and the embryonic cleft is in the process of closing (Fig. 1–3). The inner wall (retinal) of the optic cup is initially in direct continuity with the rest of the retinal layers in its most proximal part. The axons of the retinal ganglion cells extend to this area of continuity, then turn at right angles, traversing the retinal layer to reach the optic stalk.

As the axons reach the rim of the stalk, some cells are cut off from the main body and become sequestrated as a clump of glial cells in the central area of the presumptive optic disc, forming the primitive epithelial papilla of Bergmeister. The papilla becomes vascularized by the hyaloid artery, and its cells form the vascular sheath. When the main vessel disappears before birth (7 to 8 months), the papilla atrophies; the degree of atrophy ultimately determines the depth of the physiologic disc cup. Some of its cellular elements may persist in the glial sheaths of the developing retinal arteries.

The advancing nerve fibers enter the optic stalk (17 to 19 mm) and course toward the brain's inferonasal aspect. By the 19-mm stage, the lumen has been almost com-

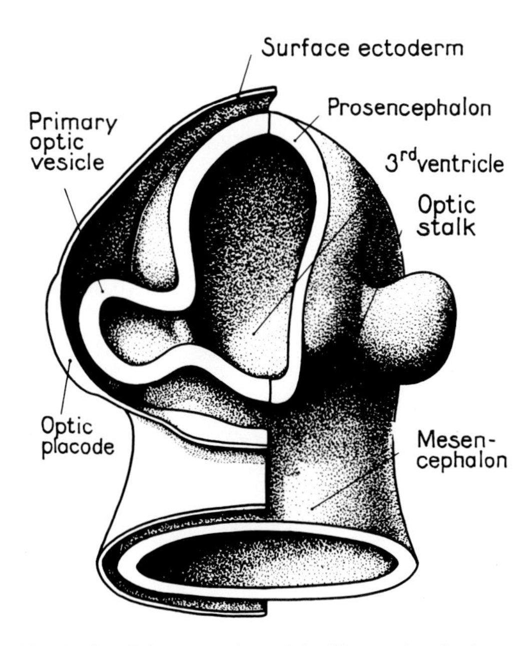

Fig. 1–1. Primary optic vesicle (4-mm stage), demonstrating open connection by the optic stalk to both the optic vesicle and the forebrain.

pletely filled in with nerve fibers, and by the 25-mm stage, the whole stalk is occupied. Some of the cells derived from the primitive epithelium of the optic stalk are transformed into supporting glial cells of the optic nerve.

As development proceeds from the 25-mm stage, the cavity of the optic vesicle no longer communicates with the forebrain cavity. The recessus opticus in the floor of the third ventricle marks the cerebral end of the original connection between the two.

Meanwhile, the distal lips of the embryonic cleft have fused (10 to 11 mm), so that the hyaloid artery, after entering the proximal end of the fissure with its accompanying mesoderm, becomes buried in the nerve fiber layer; eventually, the buried hyaloid artery becomes the intraneural portion of the central retinal artery. A graphic representation of this process appears in Figure 1–4.

As early as the 10-mm stage, the sheaths surrounding the optic nerve become apparent as layers of condensation from the surrounding tissue. A single compact coat forms by the 17-mm stage, and at the 45- to 50-mm stage, the pial sheath is defined.

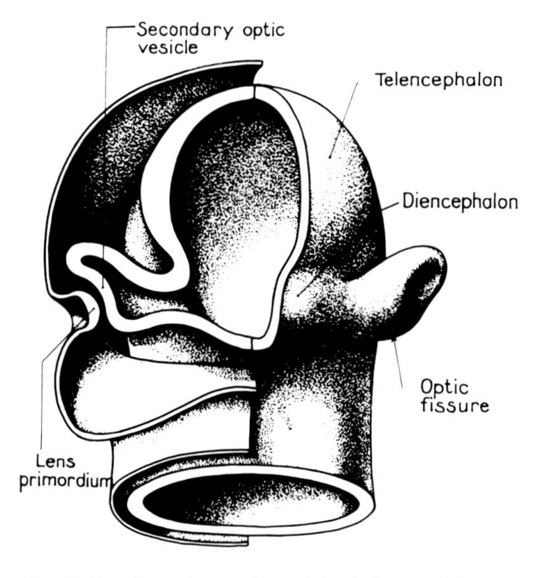

Fig. 1–2. Secondary optic vesicle during early invagination of the optic vesicle and optic stalk.

The dura mater is distinguishable by the fifth fetal month and the arachnoid becomes differentiated between it and the pial sheath by the sixth month. The pia mater and the arachnoid are largely ectodermal in origin, not mesodermal, as formerly believed.

During the eighth and ninth months, the architecture of the optic nerve is essentially complete. The last part of the framework to be fully differentiated is the lamina cribrosa. Mesodermal elements are discernible by the sixth month and well in evidence by the seventh month. During the eighth month, the lamina is permeated by collagenous fibers from the sclera, choroid, dura mater, and connective tissue from the hyaloid system. This structure is not fully consolidated until a time soon after birth.

The optic nerve continues to increase in both diameter and length for some time beyond birth. At birth, its diameter is 2.0 mm and its length 24 mm. By the third month, its diameter widens to 2.4 mm because of the increased thickness of the myelin sheaths. These dimensions of the optic nerve expand throughout much of the child's growth period, until at puberty the diameter averages 3.5 mm, and the length 45 mm.

ANATOMY

The optic nerve extends from the posterior pole of the eye to the optic chiasm, where the fibers arising from the retinal ganglion cells nasal to the fovea cross to the opposite tract and the temporal fibers continue into the ipsilateral optic tract (Fig. 1–5).

The optic nerve contains over one million visual and pupillomotor fibers. It may be described as having four segments: (1) intraocular, (2) orbital, (3) intraosseous, and (4) intracranial.

The intraocular portion of the optic nerve occupies the posterior scleral foramen. It is approximately 1 mm long and 1.5 mm in diameter at the disc; as it leaves the eye, the diameter expands to approximately 3.5 mm.

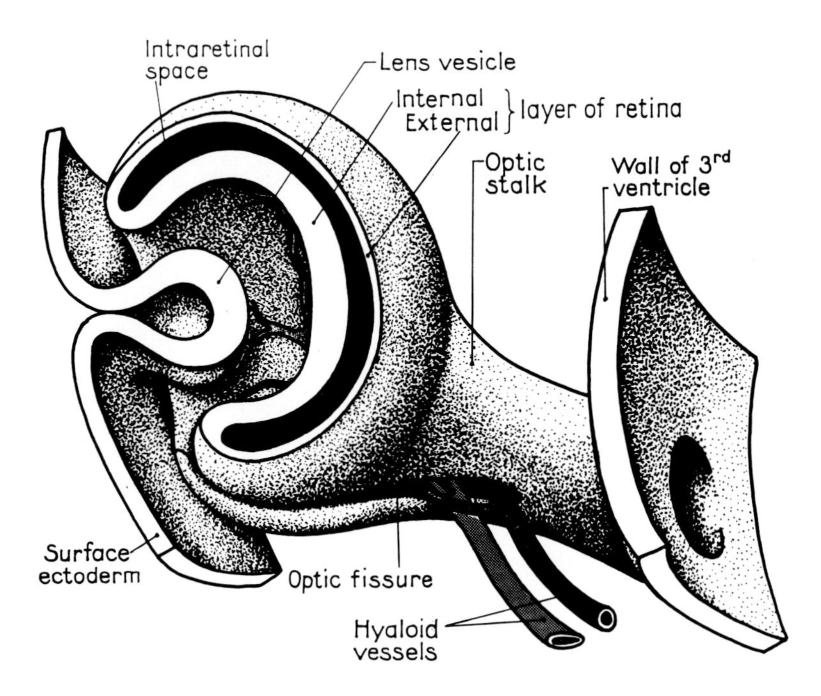

Fig. 1–3. Later optic vesicle invagination (33 days) with closing of the embryonic cleft.

The orbital portion of the nerve is from 20 to 30 mm long, and extends from the globe to the optic foramen, forming an elongated S shape that allows free movement of the globe. The nerve is surrounded by dura mater, arachnoid, and pia mater. As the nerve leaves the eye, it is surrounded by the posterior ciliary arteries. The central retinal artery and vein enter its inferomedial surface 10 to 15 mm behind the globe. At the optic foramen, the ophthalmic artery lies below and lateral to the nerve. The inferior division of the oculomotor nerve, the nasociliary artery, the abducens, and the ciliary ganglion lie just lateral to the optic nerve in the posterior orbit. At the apex of the orbit, the optic nerve is surrounded by the recti muscles that arise from the fibrous circle of Zinn.

The optic nerve then enters the optic canal through the canal's anterior opening (optic foramen) in the apex of the roof of the orbit. The canal, which represents the intraosseous portion of the optic nerve, is 5 to 8 mm long and 4 to 6 mm wide; it runs in a posteromedial course through the root of the lesser wing of the sphenoid bone. The canal transmits the optic nerve, ophthalmic artery, sympathetic branches of the carotid plexus, and the extensions of the meningeal sheaths. The sphenoid sinus occasionaly extends into the roof of the canal.

The intracranial portion of the optic nerve extends posteriorly and medially to reach the chiasm and is usually 10 to 15 mm long. The frontal lobe of the brain lies above the optic nerve. On the central surface of each frontal lobe, the olfactory tract is separated from the optic nerve by the anterior cerebral and anterior communicating arteries. On its lateral side, the optic nerve forms an immediate relationship with the carotid artery, as the artery emerges from the cavernous sinus. The internal carotid artery is actually attached to the optic nerve by the ophthalmic artery, which arises from the carotid artery and lies within the dural sheath of the optic nerve. Inferiorly, the relationship to the sphenoid sinus is important anatomically.

The optic nerve is surrounded by the same sheaths that surround the brain. The optic

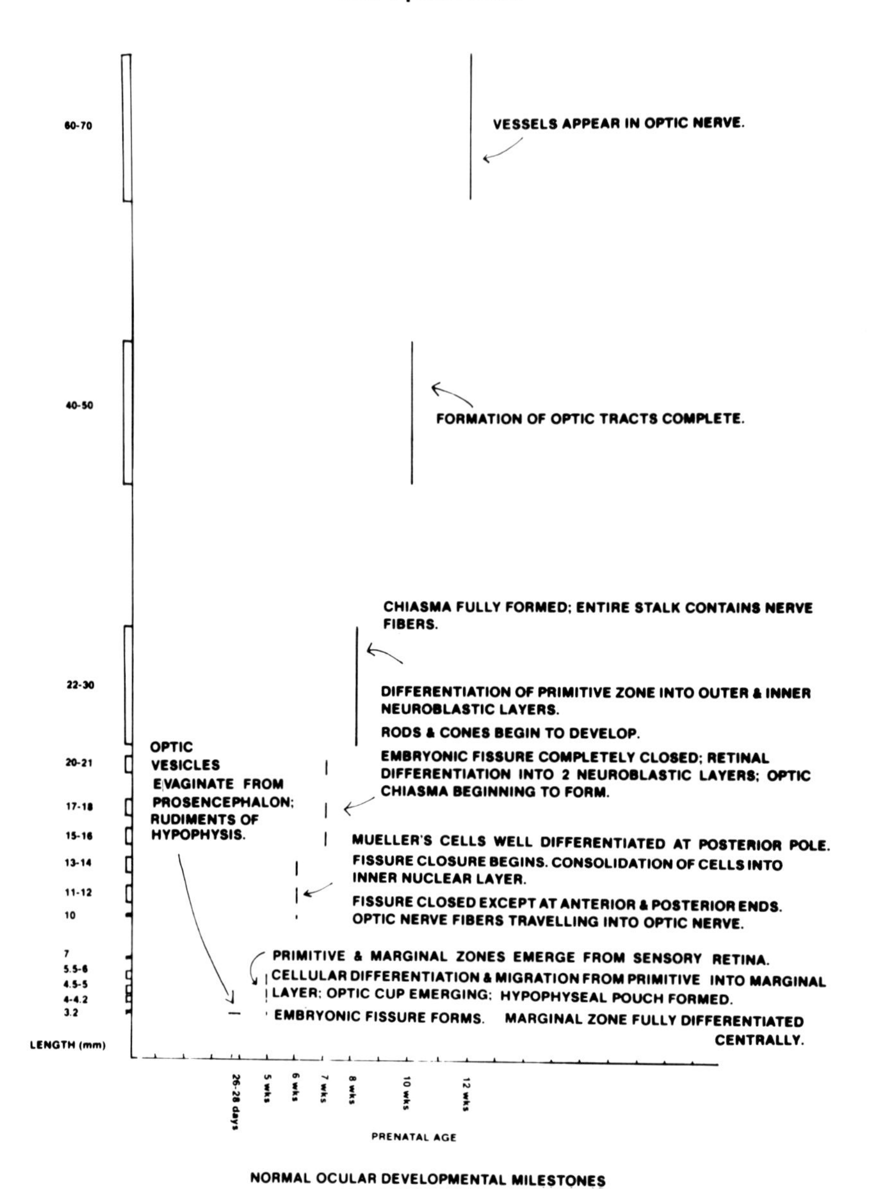

Fig. 1–4. Normal ocular developmental milestones.

nerve's dura mater is continuous with the brain's dura mater and merges anteriorly with the sclera. It is not continuous with the optic nerve but is joined to the nerve in the optic canal. At its interior exit from the optic canal, the dura mater splits into two layers, the outer layer forming the periorbital fascia and the inner layer continuing forward around the nerve to fuse with the anterior layer of the sclera. The dura mater surrounding the optic nerve receives blood vessels from the ophthalmic artery. The subdural space about the nerve is continuous with the subdural space about the brain; however, this space may be more potential than real.

The arachnoid sheath forms the middle covering of the optic nerve and is continuous posteriorly with the arachnoid of the brain. The subarachnoid space is real and

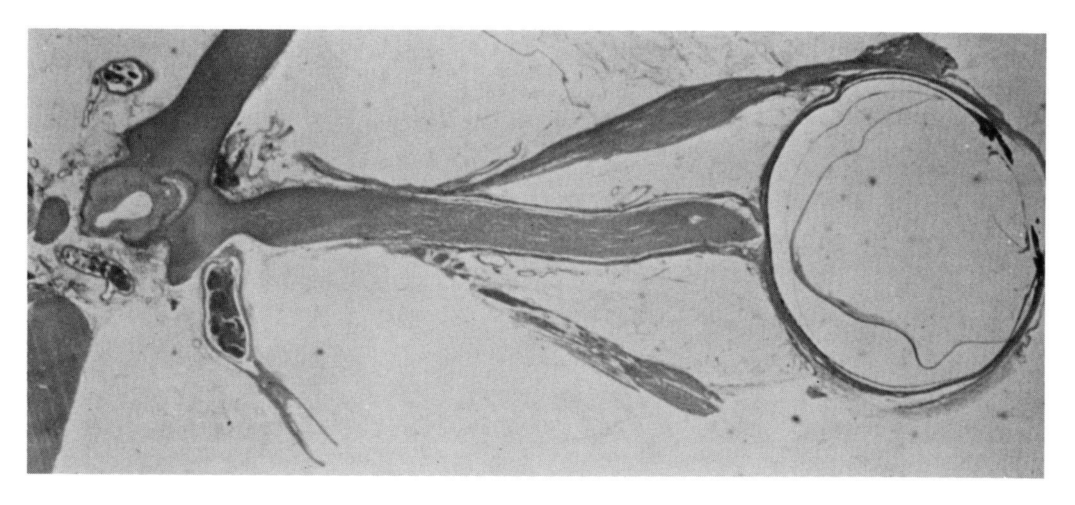

Fig. 1–5. Anatomic section of the optic nerve, demonstrating its entire course from the globe to the chiasm.

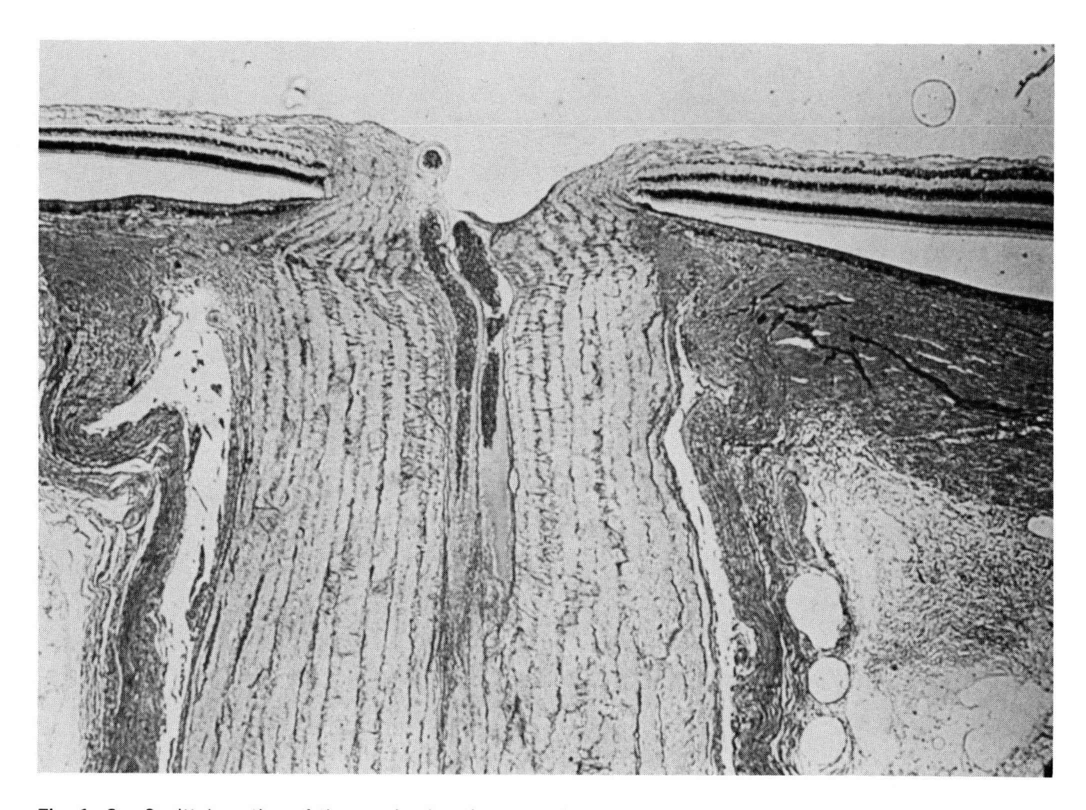

Fig. 1–6. Sagittal section of the proximal optic nerve, demonstrating the entering retinal fibers (ganglion cell axones), intraneural septae, central vessels, and meningeal sheaths, Note the anterior merging of the dural layer with the sclera.

continuous. The arachnoid contains trabeculae that originate from the dural layer and extend to the pia mater, with which they unite. The arachnoid itself is almost devoid of blood vessels.

The pial sheath is in immediate contact with the nerve. It joins the sclera anteriorly and continues with the brain's pia mater posteriorly. From the inner aspect of the pia mater, septa extend into the nerve and divide the nerve fibers into bundles. The pia mater receives a rich blood supply from the ophthalmic artery.

The principal blood supply for the trunk of the optic nerve is derived from the pial sheath, whose blood supply is derived in turn from the ophthalmic artery. There occasionally may be a recurrent artery that arises from the central retinal artery and supplies blood to the core of the nerve. There also may be a central optic nerve artery that arises from the ophthalmic artery and is independent of the central retinal artery. As the nerve enters the globe anteriorly, it derives additional vascular supply from the short ciliary arteries and the vascular circle of Haller-Zinn.

The normal anatomy of the optic nerve, with its meningeal sheaths, pial septae, and central vessels, is depicted in Figure 1–6.

Anderson, D.R.: Ultrastructure of human and monkey lamina cribrosa and optic nerve head. Arch. Ophthalmol., *82*:800–814, 1969.

Barber, A.N.: Embryology of the Human Eye. St. Louis, C.V. Mosby, 1955.

Barber, A.N., Ronstrom, G.N., and Muelling, R.G., Jr.: Development of the visual pathway: Optic chiasm. Arch. Ophthalmol., *52*:447–453, 1954.

Duke-Elder, S.: Embryology. *In* System of Ophthalmology. Vol. 3. St. Louis, C.V. Mosby, 1963, Pt. 1.

Francois, J.: Heredity in Ophthalmology. St. Louis, C.V. Mosby, 1961.

Hayreh, S.S.: Anatomy and physiology of the optic nerve head. Trans. Am. Acad. Ophthalmol. Otolaryngol., *78*:240–254, 1974.

Magoon, E.H., and Rabb, R.M.: Development of myelin in human optic nerve and tract. Arch. Ophthalmol., *99*:655–659, 1981.

Mann, I.: Developmental Abnormalities of the Eye. Cambridge, University Press, 1937.

Radius, R.L, and Gonzales, M.: Anatomy of the lamina cribrosa in human eyes. Arch. Ophthalmol., *99*:2159–2162, 1981.

Ranson, S.W.: The Anatomy of the Nervous System: Its Developmental Function. 7th Ed. Philadelphia, W.B. Saunders, 1943.

Reeh, M., Wobig, J., and Wirtschafter, J.: Ophthalmic Anatomy. Continuing Medical Education Program, San Francisco, American Academy of Ophthalmology, 1981.

Tuchmann-Duplessis, H., Auroux, M., and Haegel, P.: Illustrated Human Embryology. Vol. 3, Nervous System and Endocrine Glands. New York, Springer-Verlag. 1975.

Waardenburg, P.J., Franceschetti, A., and Klein, D.: Genetics and Ophthalmology. Vol. 1. Springfield, Ill., Charles C Thomas, 1962.

Walsh, F.B., and Hoyt, W.F.: Clinical Neuro-Ophthalmology. 3rd Ed. Vol. 1. Baltimore, Williams & Wilkins, 1969.

Wise, G.N., Dollery, C.T., and Henkind, P.: The Retinal Circulation. New York, Harper and Row, 1971.

Wolff, E.: Anatomy of the Eye and Orbit. Philadelphia, W. B. Saunders, 1961.

Chapter 2

OPTIC NERVE ANOMALIES

Congenital Optic Atrophy

Precise differentiation between true congenital optic atrophy and optic atrophy with an infantile onset is often difficult, if not impossible. Debate exists as to whether these optic nerves are truly "atrophic" or "dystrophic," that is, whether the nerves developed and subsequently atrophied or were actually malformed from the start. The pattern of inheritance can vary from autosomal recessive to dominant with regular to irregular penetrance, or the optic atrophy may present as a sporadic occurrence (Table 2–1).

HEREDITARY OPTIC ATROPHY

Recessive congenital optic atrophy is characterized by pronounced visual deficiency, nystagmus, color vision defects, and bilateral pallor of the optic discs. The retinal vessels may also be narrowed. The electroretinogram is usually normal, and the visual evoked response is reduced or extinguished.

Dominant congenital optic atrophy has been described most commonly in males, with the inheritance dominant throughout. Except for pallor of the discs, there are no fundus changes. Reduced vision and nystagmus are usually present in all affected individuals. The electroretinogram is usually normal; the visual evoked response is markedly reduced.

In true congenital optic atrophy, the nerve fibers formed but subsequently atrophied (Fig. 2–1). The typical case remains static, neither deteriorating nor improving. The optic atrophy associated with disturbances of the central nervous system (Behr's hereditary optic atrophy) is not congenital, but starts in infancy. Leber's optic atrophy usually does not become apparent until after puberty.

Congenital optic atrophy is usually distinguishable from optic nerve hypoplasia, in which the optic nerve is not only pale but also one third to two thirds the normal size. Hereditary or familial occurrence is rare in optic nerve hypoplasia.

Associated Anomalies

Optic atrophy occurring at birth, or noted soon after, has been found in association with most types of the craniosynostoses and craniofacial dysraphic states, including:

(1) Cranium bifidum
(2) Spina bifida
(3) Dysostosis craniofacialis (Crouzon's disease)
(4) Hypertelorism
(5) Microcephaly

TABLE 2–1. HEREDOFAMILIAL OPTIC ATROPHIES

	DOMINANT	RECESSIVE			INDETERMINATE
	Juvenile (Infantile)	Early infantile (Congenital), Simple	Behr's type Complicated	With diabetes mellitus, ± deafness	Leber's Disease
Age at onset	Childhood (4 to 8 years)	Early childhood (3 to 4 years)	Childhood (1 to 9 years)	Childhood (6 to 14 years)	Early adulthood (18 to 30 years; up to sixth decade)
Visual impairment	Mild/moderate (20/40–20/200)	Severe (20/200-HM)	Moderate (20/200)	Severe (20/400-FC)	Moderate/severe (20/200-FC)
Nystagmus	Rare	Usual	In 50%	Absent	Absent
Optic disc	Mild temporal pallor; ± temporal excavation	Marked diffuse pallor (± arteriolar attenuation)	Mild temporal pallor	Marked diffuse pallor	Moderate diffuse pallor; disc swelling in acute phase
Color vision	Blue-yellow dyschromatopsia	Severe dyschromatopsia/ achromatopsia	Moderate to severe dyschromatopsia	Severe dyschromatopsia	Dense central scotoma for colors
Course	Variable, slight progression	Stable	Stable	Progressive	Acute visual loss, then usually stable; may improve/worsen

From Glaser, J.S.: Heredofamilial disorders of the optic nerve. *In* Genetic and Metabolic Eye Disease. Edited by M.F. Goldberg. Boston, Little, Brown, 1974.

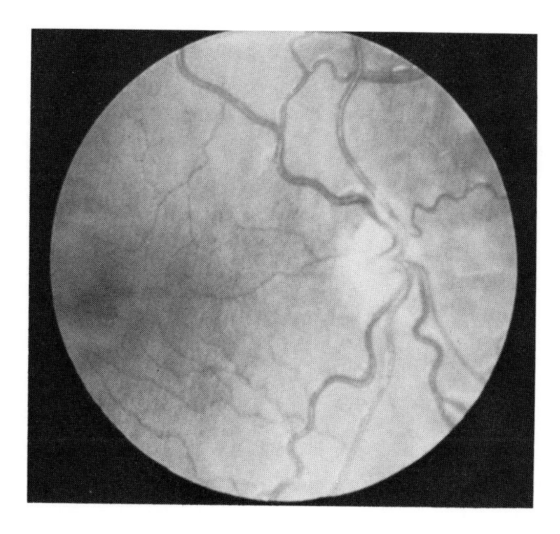

Fig. 2–1. Congenital optic atrophy with pallor of the disc, normal vascular pattern, and indistinct macular reflex.

(6) Oxycephaly
(7) Platybasia
(8) Scaphocephaly

It has also been described in association with:

(1) Congenital spastic diplegia (Little's disease)
(2) Hereditary cerebellar ataxia
(3) Spinocerebellar ataxia
(4) Congenital mental deficiency
(5) Chondrodystrophy
(6) Osteopetrosis (Albers-Schönberg disease)
(7) Fibrous dysplasia of the skull
(8) Chondrodysplasia (Ollier's disease)
(9) Mucopolysaccharidosis
(10) Neurofibromatosis (von Recklinghausen's disease)
(11) Hand-Schüller-Christian disease
(12) Encephalotrigeminal angiomatosis (Sturge-Weber syndrome)

BIRTH INJURIES

Unilateral and, rarely, bilateral optic atrophy may result from birth injury, from direct trauma to the optic nerve through distortion or traction, or from fracture of the orbit in the area of the optic foramen.

In cases of difficult forceps delivery, fracture of the orbit or tearing of the tentorium may occur and the resulting intracranial hemorrhage can be fatal. Subdural hematomas of the optic nerve may also cause optic atrophy in infants. Similarly, intracranial subarachnoid hemorrhage may occur, traversing the subarachnoid space of the optic nerve and extending even into the vitreous cavity.

Optic atrophy may also arise from dislocation of the eye during delivery, which may result in stretching or contusion of the nerve and, in rare cases, actual avulsion of the nerve. External evidence of trauma is usually apparent.

Neonatal asphyxia rarely produces optic atrophy, but when atrophy does occur, it is usually bilateral and associated with other evidence of cerebral hypoxia.

Injury, hypoxia, or dysgenesis of the immature fetal visual cortex may give rise to trans-synaptic neuronal degeneration with atrophy of the infrageniculate pathways and subsequent atrophy of the optic nerve. This is associated most often with severe, diffuse cerebral dysfunction, cerebral palsy, hemiparesis, seizures, and mental retardation.

CEREBRORETINAL ABIOTROPHIES

A disparate group of central nervous diseases known as the cerebroretinal abiotrophies are characterized by congenital or infantile visual defects, seizures, mental retardation, and a variety of other cerebral, visceral, retinal, and optic nerve dystrophies (Table 2–2).

Lipid Storage Diseases

The cerebroretinal lipid storage diseases (lipidoses) may occur in the form of congenital or infantile blindness with optic nerve atrophy, failure of normal development, and diffuse neurologic dysfunction.

Congenital familial amaurotic idiocy and the infantile form of Tay-Sachs disease represent cerebroside abnormalities, probably caused by the deficiency of the enzyme hexosaminidase A. Tay-Sachs disease, which

TABLE 2–2. DIFFERENTIAL DIAGNOSIS OF CEREBRAL RETINAL DYSTROPHIES

	AGE OF PRESENTATION	DISTRIBUTION	CLINICAL	LABORATORY
Tay-Sachs disease	1 mo–1 yr	Jewish	CNS signs	Serum, leukocytes (hexosaminidase A)
Sandhoff's disease	1 mo-1 yr	Multiracial	CNS signs $\bar{c}$ mild or no visceromegaly	Serum, leukocytes (total hexosaminidase)
GM_1-gangliosidosis	2 mo–5 yr	Multiracial	CNS signs $\bar{c}$ visceromegaly	Leukocytes, liver, skin (β-galactosidase)
Niemann-Pick	3 mo	Jewish and Multiracial	CNS signs $\bar{c}$ visceromegaly	Leukocytes (sphingomyelinase)
Mucolipidosis I	3 mo	Multiracial	CNS signs $\bar{c}$ mild or no visceromegaly	Normal or elevated liver acid hydrolases
Metachromatic Leukodystrophy	3 mo	Multiracial	CNS signs	Leukocytes, liver (sulfatase)

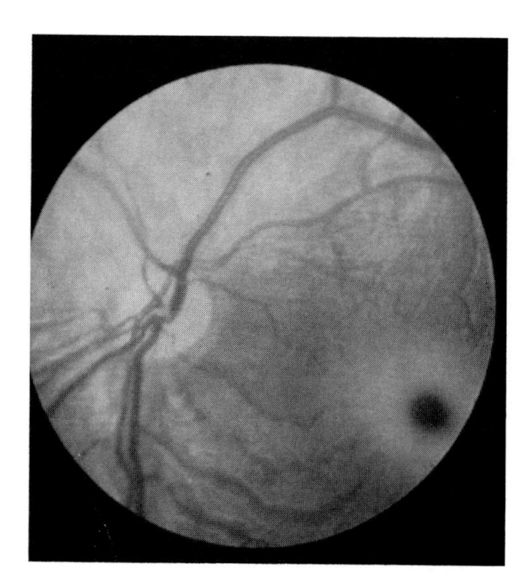

Fig. 2–2. Tay-Sachs disease. The cherry red spot appears in the macula with diffuse pallor of the optic nerve head.

has a recessive hereditary pattern, occurs almost exclusively in infants of Jewish parentage.

The cherry red spot in the macula and pallor of the optic nerves are common and are caused by abnormal storage of GM_2-ganglioside in the gangliion cells (Fig. 2–2).

Metachromatic Leukodystrophies

These form a related group of demyelinizing diseases with a sulfatide-sulfatase enzyme deficiency. They appear between birth and three years of age with optic atrophy and diffuse central nervous system involvement. The entire spectrum of these and other inborn enzyme deficiencies is summarized in Table 2–3.

Leber's Congenital Amaurosis

Leber's congenital amaurosis, a recessive form of congenital or infantile tapetoretinal dystrophy, is characterized by blindness, mental retardation, seizures, and other neurologic disorders. The ocular fundi may appear quite normal early on, but may later develop pigmentary degeneration of the retina and pallor of the optic nerves. Electroretinography reveals a virtually extinguished response.

Congenital Stationary Cone Dystrophy

Another congenital retinal abiotrophy is congenital stationary cone dystrophy, or

TABLE 2–3. INBORN LYSOSOMAL DISORDERS WITH OPTIC ATROPHY

DISEASE	ENZYME DEFICIENCY
MUCOPOLYSACCHARIDOSES	
Hurler (I–H)	α-L-iduronidase
Scheie (I–S)	α-L-iduronidase (partial)
Hunter A (IIA) (severe phenotype)	Iduronate sulfatase
Hunter B (IIB) (mild phenotype)	Iduronate sulfatase
San Filippo A (IIIA)	Heparan sulfate sulfatase
San Filippo B (IIIB)	N-acetyl-glucosaminidase
Morquio (IV)	N-acetyl-galactosamine sulfatase
Maroteaux-Lamy A (VIA) (severe phenotype)	Arylsulfatase B
LIPIDOSES	
Generalized gangliosidosis (GM_1-I)	β-galactosidase A,B,C
Juvenile gangliosidosis (GM_1-II)	β-galactosidase B,C
Tay-Sachs (GM_2-I)	Hexosaminidase A
Sandhoff (GM_2-II)	Hexosaminidase A,B
Juvenile GM_2 gangliosidosis (GM_2-III)	Hexosaminidase A (partial)
Niemann-Pick A (infantile)	Sphingomyelinase
Krabbe (globoid cell leukodystrophy)	Galactocerebroside β-galactosidase
Metachromatic leukodystrophy (infantile)	Arylsulfatase A
Metachromatic leukodystrophy (juvenile and adult)	Arylsulfatase A
Sulfatidosis, Austin variant (mucosulfatidosis)	Arylsulfatase A,B,C

"achromatopsia," which primarily affects the retinal cones. The disease occurs with reduced visual acuity, color blindness, and photophobia. Pendular nystagmus, subalbinotic appearance of the fundus, and pallor of the optic nerves usually occur as associated signs of the disease. Electroretinography demonstrates the photopic defect.

Congenital Oculo-Acoustico-Cerebral Dystrophy

This disease, also called Norrie's Disease, is an X-linked hereditary pattern dystrophy characterized by reduced visual acuity, mental retardation, deafness, and optic atrophy. It may also be associated with persistent hyperplastic vitreous, cataract, and corneal opacification.

RELATED TERMS

(1) Autosomal dominant optic atrophy
(2) Behr's optic atrophy
(3) Congenital familial amaurotic idiocy
(4) Congenital stationary cone dystrophy (achromatopsia)
(5) Congenital/infantile optic atrophy
(6) Ganglion cell infiltration disease
(7) Gangliosidosis
(8) Leber's hereditary optic atrophy (congenital amaurosis)
(9) Metachromatic leukodystrophies
(10) Oculo-acoustico-cerebral dystrophy (Norrie's disease)
(11) Sandhoff's disease
(12) Tay-Sachs disease

Allen, R., McCusker, J.J., and Tourtellote, W.W.: Metachromatic leukodystrophies. Pediatrics, *30*:629, 1962.

Alstrom, C.H., and Olson, O.: Heredo-retinopathia congenitalis. Hereditas, *42*:1–178, 1957.

Apple, D.J., and Rabb, M.F.: Clinicopathologic Correlation of Ocular Disease: A Text and Stereoscopic Atlas. St. Louis, C.V. Mosby, 1978.

Aronson, S.M.: Infantile amaurotic infancy. Pediatrics, *26*:229–242, 1960.

Austin, J., et al.: Metachromatic leukodystrophy. Arch. Neurol., *18*:225–240, 1968.

Behr, C.: Die komplizierte, heriditar-familiare optekusatrophie des kindesalters. Klin. Monatsbl. Augenheilkd., *46*:138–169, 1909.

Budka, H., Seemann, D., and Danielczyk, W.: Hereditary cerebellar atrophy (Holmes type) with optic atrophy: A clinico-pathological study of four generations in a family. Arch. Psychiatr. Nervenkr., *226*:311–318, 1979.

Cagianut, B., et al.: Thiosulphate-sulphur transferase (Rhodanese) deficiency in Leber's hereditary optic atrophy. Lancet, *2*:981–982, 1981.

Dorfman, L.J., et al.: Visual evoked potentials in Leber's hereditary optic neuropathy. Ann. Neurol., *1*:565–568, 1977.

Duke-Elder, S.: Congenital optic atrophy. *In* System of Ophthalmology. Vol. 3. St. Louis, C.V. Mosby, 1963, Pt. 2.

Ferrier, P.E., et al.: Cerebral gigantism (Soto's Syndrome) with juvenile macular degeneration. Helv. Paediatr. Acta, *35*:97–102, 1980.

Francois, J.: Heredity in Ophthalmology. St. Louis, C.V. Mosby, 1961.

Glaser, J.S.: The heredodegenerative optic atrophies. In Clinical Ophthalmology. Vol. 2. Edited by I.D. Duane. Philadelphia, Harper and Row, 1981.

Goldberg, M.F.: Genetic and Metabolic Eye Disease. Boston, Little, Brown, 1974.

Green, J.B.: Cerebral lipidoses. *In* Handbook of Clinical Neurology. Vol. 15, The Epilepsies. Edited by P.J. Vinken and G.W. Bruyn. New York, American Elsevier, 1974.

Griscom, J.M.: Hereditary optic atrophy. Am. J. Ophthalmol., *4*:347–352, 1921.

Harding, G.F., and Crews, S.J.: The visual evoked potential in hereditary optic atrophy. Adv. Neurol., *32*:21–30, 1982.

Harding, G.F., Crews, S.J., and Pitts, S.M.: Psychophysical and visual evoked potential findings in hereditary optic atrophy. Trans. Ophthalmol. Soc. U.K., *99*:96–102, 1979.

Kjer, P.: Infantile optic atrophy with dominant mode of inheritance. Acta Ophthalmol. [Suppl.] (Copenh.), *54*:9–146, 1959.

Kollarits, C.R., et al.: The autosomal dominant syndrome of progressive optic atrophy and congenital deafness. Am. J. Ophthalmol., *87*:789–792, 1979.

Levine, R.A., Rosenberg, M.A., and Rabb, M.F.: Optic atrophy (hereditary forms). *In* Principles and Practice of Ophthalmology. Vol. 3. Edited by G.A. Peyman, D.R. Saunders, and M.F. Goldberg. Philadelphia, W.B. Saunders, 1980.

Mullaney, J.: Normal development and developmental anomalies of the eye. *In* Pathobiology of Ocular Disease: A Dynamic Approach. Edited by A. Garner and G.K. Klintworth. New York, Marcel Dekker, 1982.

Newman, W.: Congenital optic atrophy. Roy. Lond. Ophthalmol. Hosp. Rep., *4*:202, 1864.

Stanescu, B., et al.: Electroretinographic changes in a case of spino-cerebellar degeneration (SCD). Metab. Pediatr. Ophthalmol., *4*:221–223, 1980.

Stendahl-Brodin, L., Möller, E., and Link, H.: Hereditary optic atrophy with probable association with a specific HLA haplotype. J. Neurol. Sci., *38*:11–21, 1978.

Trobe, J.D., Glaser, J.S., and Cassady, J.C.: Optic

atrophy. Differential diagnosis by fundus observation alone. Arch. Ophthalmol., *98*:1040–1045, 1980.
Van Lith, G.H.: Difficulties in diagnosing Leber's optic atrophy. Doc. Ophthalmol., *48*:255–259, 1980
Waardenburg, P.J.: Different types of hereditary optic atrophy. Acta Genetica Et Statistica Medica, *7*:287–290, 1957.
Westmoreland, B.F., Groover, R.V., and Sharbrough, F.W.: Electrographic findings in three types of cerebromacular degeneration. Mayo Clin. Proc., *54*:12–21, 1979.
Williams, T.D.: Annual review of section on ocular disease: Diseases of the optic nerve, tracts, and visual cortex. Am. J. Optom. Physiol. Opt., *57*:33–47, 1980.

Congenital Pigmentation of the Optic Disc

Pigmentation of the optic disc may occur in several forms: (1) pigment flecks on the surface of the nerve head or in the lamina cribrosa, (2) dense pigment plaques either on or extending outward from the disc, (3) linear stripes of pigment, and (4) a slate-gray color of the entire nerve heads.

The first three types are most likely caused by developmental inclusion of pigment-bearing tissues in the lamina cribrosa or by extension of pigment epithelium into and across the disc margin. Cellular metaplasia in the developing optic stalk may also occur (Fig. 2–3).

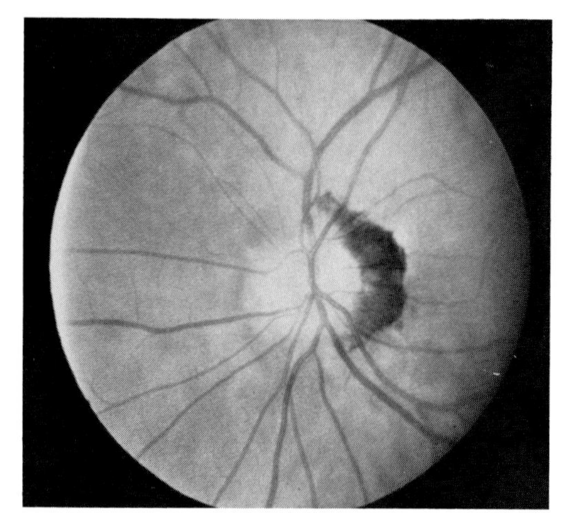

Fig. 2–3. Pigmented crescent in the form of a dense pigment plaque extending from the temporal disc.

The fourth type of congenital pigmentation, a slate-gray color of the entire disc, is of specific interest (Fig. 2–4). Several reports in the literature associate this peculiar disc pigmentation with "delayed visual development." The described infants had appeared to be blind in the neonatal period, but within months to years later, they demonstrated visual attention and function, more active pupillary responses, and discs of practically normal color. The cause of this phenomenon is unknown but has been attributed to delayed myelinization of the optic nerves. This condition has also been referred to as "gray pseudo-optic atrophy" and as "myelogenous dysgenesis."

As an anecdotal observation, I have examined several infants with grayish pigmentation of the optic nerves in association with hypoplastic discs, who seemed to develop progressive improvement in "visual attention" over a two- to three-year period.

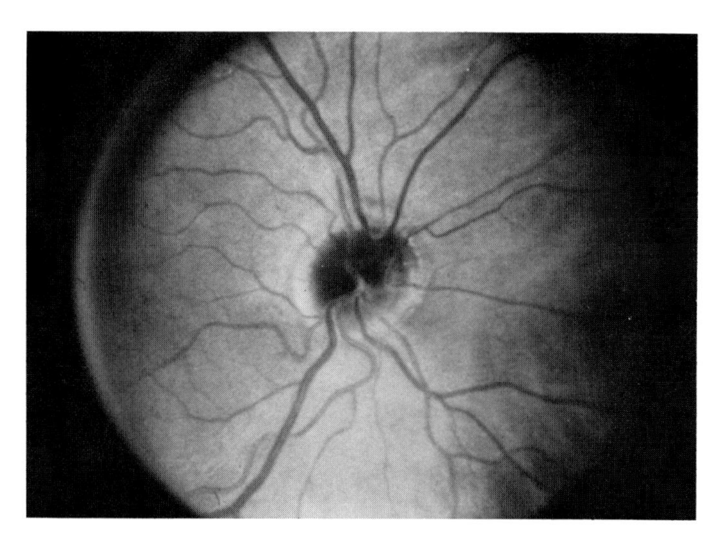

Fig. 2–4. Congenital pigmentation of the entire disc, perhaps a result of delayed myelinization.

This same phenomenon has been seen in albino neonates, with apparent improvement in visual attention occurring at about the same time as the nystagmus first appears!

RELATED TERMS

(1) Congenital pigmentation of the optic nerve
(2) Delayed myelinization of the optic nerve
(3) Gray pseudo-optic atrophy
(4) Myelogenous dysgenesis

Beauvieux, J.: La pseudo-atrophie optique des nouveau-new. Ann. Oculist., *163*:881–921, 1926.

Blodi, F.C., et al.: Pseudomelanocytoma of the optic nerve head. Arch. Ophthalmol., *73*:353–355, 1965.

Coats, G.: Congenital pigmentation of the papilla. Roy. London Ophthalmic Hosp. Rep., *17*:225–231, 1908.

Duke-Elder, S.: Congenital optic atrophy. *In* System of Ophthalmology. Vol. 3. St. Louis, C.V. Mosby, 1963, Pt. 2.

Halbertsma, K.T.: Pseudo-atrophy of the optic nerve. Ned. Tijdschr. Geneesk., *81*:1230–1236, 1937.

Je Zak-Lipska, A.: Slate-coloured optic nerve discs. Klin. Oczna., *81*:581–582, 1979.

Juarez, C.P., and Tso, M.O.: An ultrastructural study of melanocytomas (magnocellular nervi) of the optic disk and uvea. Am. J. Ophthalmol., *90*:48–62, 1980.

Kawamura, S., et al.: Orientational changes of the transition dipole monent of retinal chromophore on the disk membrane due to the conversion of rhodopsin to bathorhodopsin and to isorhodopsin. Vision Res., *19*:879–884, 1979.

Kravitz, D.: Pigmentation of the optic disk. Arch. Ophthalmol., *29*:826–830, 1943.

Mann, I.: Developmental Abnormalities of the Eye. Cambridge, University Press, 1937.

Reese, A.B.: Pigmentation of the optic nerve. Arch. Ophthalmol., *9*:560–570, 1933.

Rosenthal, A.R., Falconer, D.G., and Barrett, P.: Digital measurement of pallor-disc ratio. Arch. Ophthalmol., *98*:2027–2031, 1980.

Shields, M.B.: Grey crescent in the optic nerve head. Am. J. Ophthalmol., *89*:238–244, 1980.

Sobanski, J.: Eineschiefergrave sehnervenpapille. Klin. Monatsbl. Augenheilkd., *102*:704–706, 1939.

Sorensen, P.N.: The colour of the optic disc variation with location of illumination. Acta Ophthalmol. (Copenh.), *58*:1005–1010, 1980.

Medullated Nerve Fibers

In the fetus, myelinization of the optic nerve fibers proceeds peripherally from the lateral geniculate body, and by full term, sheaths have reached the lamina cribrosa of the optic nerve head.

The explanation for their occasional further extension through the lamina cribrosa and into the retina is unknown. The abnormal presence of oligodendroglia in the retinal nerve fibers may be responsible for their "eccentric" development.

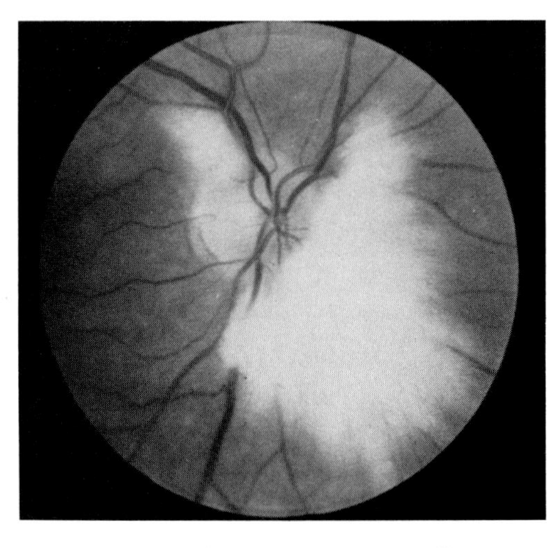

Fig. 2–5. Peripapillary medullated nerve fibers.

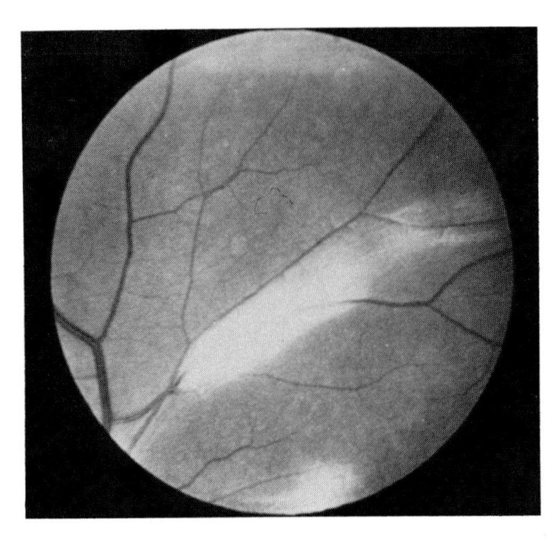

Fig. 2–6. Remote medullated nerve fibers.

These medullated fibers usually form patches that extend from the disc into the retina, and end in irregular flame-shaped fans (Fig. 2–5). Occasionally, these patches surround the disc completely or lie in the midperiphery unconnected to the disc (Fig. 2–6).

Medullated nerve fibers occur in 0.4% of all patients and are bilateral in 20% of cases. They do not usually affect visual acuity, but they may cause relative scotomas corresponding to the area of retinal involvement. The macular area is rarely involved, and the fibers usually remain unchanged throughout life.

After the lower (retinogeniculate) visual pathway is completed (48-mm stage), medullation of the nerve fibers occurs in the reverse direction, that is, centrifugally. The gradual process of myelin development depends on the glial oligodendrocytes and begins at about the fifth month in the geniculate bodies. It reaches the chiasm during the sixth and seventh months and is evident in the optic nerve in the eighth month. Its slow, distal extension reaches the level of the lamina cribrosa at birth or some several weeks thereafter, at which point the process normally ceases.

Anderson, B.: Medullated nerve fibers. Trans. Ophthalmol. Soc. U.K., *63*:343, 1942.

Barber, A.N.: Embryology of the Human Eye. St. Louis, C.V. Mosby, 1955.

Berliner, M.L.: Medullated nerve fibers associated with choroiditis. Arch. Ophthalmol., *6*:404–413, 1931.

Duke-Elder, S.: Medullated nerve fibers. *In* System of Ophthalmology. Vol. 3. St. Louis, C.V. Mosby, 1963, Pt. 2.

———: Myelination. *In* System of Ophthalmology, Vol. 3. St. Louis, C.V. Mosby, 1963, Pt. 1.

Holland, P.M., and Anderson, B., Jr.: Myelinated nerve fibers and severe myopia. Am. J. Ophthalmol., *81*:597, 1976.

Levy, N.S., and Ernest, J.T.: Retinal medullated nerve fibers. Arch. Ophthalmol., *91*:330, 1974.

Magoon, E.H., and Rabb, R.M.: Development of myelin in human optic nerve and tract. Arch. Ophthalmol., *99*:655–659, 1981.

Mann, I.: Developmental Abnormalities of the Eye. Cambridge, University Press, 1937.

Schachat, S.P., and Miller, N.R.: Atrophy of myelinated retinal nerve fibers after acute optic neuropathy. Am. J. Ophthalmol., *92*:854–856, 1981.
Sharpe, J.A., and Sanders, M.D.: Atrophy of myelinated nerve fibers in the retina in optic neuritis. Br. J. Ophthalmol., *59*:229, 1975.
Straatsma, B.R., et al.: Myelinated retinal nerve fibers. Am. J. Ophthalmol., *91*:25–38, 1981.
Straatsma, B.R., et al.: Myelinated retinal fibers. Clinicopathological study and clinical correlations. *In* XXIII Concilium Ophthalmologicum. Edited by K. Shimizu and J.A. Oosterhuis. Vol. 1. Kyoto, Excerpta Medica, 1979, pp. 693–701.
Straatsma, B.R., et al.: Myelinated retinal nerve fibers associated with ipsilateral myopia, amblyopia and strabismus. Am. J. Ophthalmol., *88*:506, 1979.

Hyaloid System Remnants

Many common developmental anomalies arise from persistence of unabsorbed tissues of the hyaloid vascular system. These anomalies have been classified as follows:

(1) Glial strands on the disc
(2) Disc membranes
(3) Disc cysts
(4) Massive connective tissue on the disc (pseudoglioma)
(5) Rudimentary strands extending into the vitreous
(6) Strands on the disc and posterior lens capsule
(7) Strands extending from the disc to the lens
(8) Similar strands containing blood
(9) Strands attached only to the lens
(10) Posterior capsular cataract
(11) Posterior lens capsule striae
(12) Persistent canal of Cloquet

It is not uncommon for some portion of the transient hyaloid vascular system to persist, as either patent blood vessels or strands of glial tissue (Fig. 2–7). Posterior disc remnants reportedly occur in 95% of premature infants and in 3% of full-term infants.

Perhaps the most common clinical observation of this kind is Bergmeister's papilla (Fig. 2–8). Its fibrous glial tissue, which is present in varying amounts on the nerve head, represents remnants of the papillary cone and invaginated mesoderm. The tissue may appear as solid masses, delicate strands, or membranous veils (Fig. 2–9).

The embryonic hyaloid system and the tunica vasculosa lentis constitute a complex vascular system that invades the developing eye posteriorly as well as anteriorly. It is essential to the developing eye, but transitory. The hyaloid artery originates as a major branch of the primitive dorsal ophthalmic artery, enters the proximal groove in the optic vesicle (5-mm stage), and rapidly advances to reach the posterior lens vesicle (6- to 7-mm stage). During this early development, the artery is in free communication with the developing choroidal vascular network posteriorly, but as the fissure closes, this anastomosis is progressively obliterated. The main vessel enters the optic stalk and emerges into the optic cup in the region of the future optic disc. It grows forward through the primary vitreous to meet with vasoformative me-

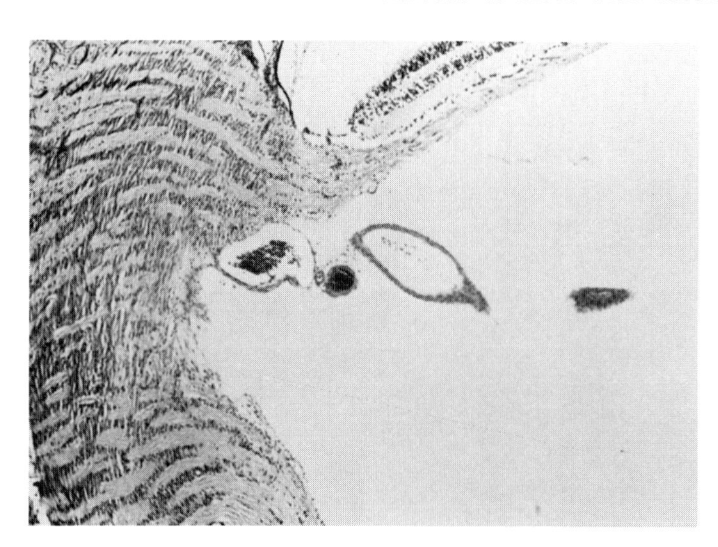

Fig. 2–7. Hyaloid system remnants extending forward from the optic disc into the posterior vitreous.

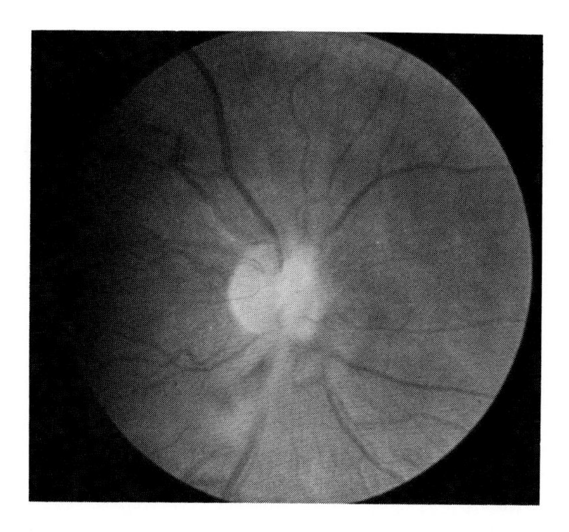

Fig. 2–8. Bergmeister's papilla extending from and obscuring central cup of nerve head.

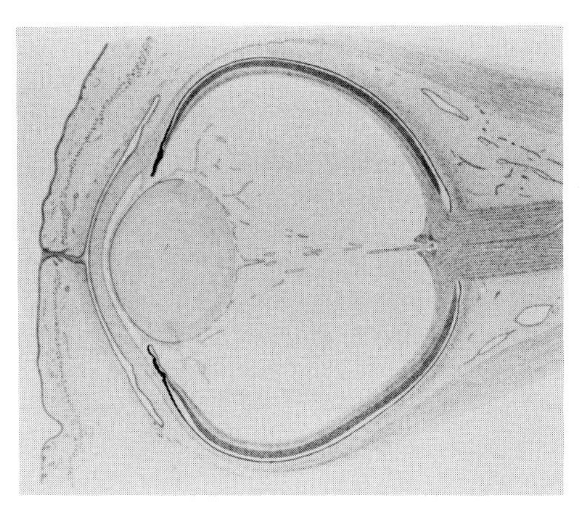

Fig. 2–10. Reabsorption of the fetal hyaloid vascular system at approximately 8 months.

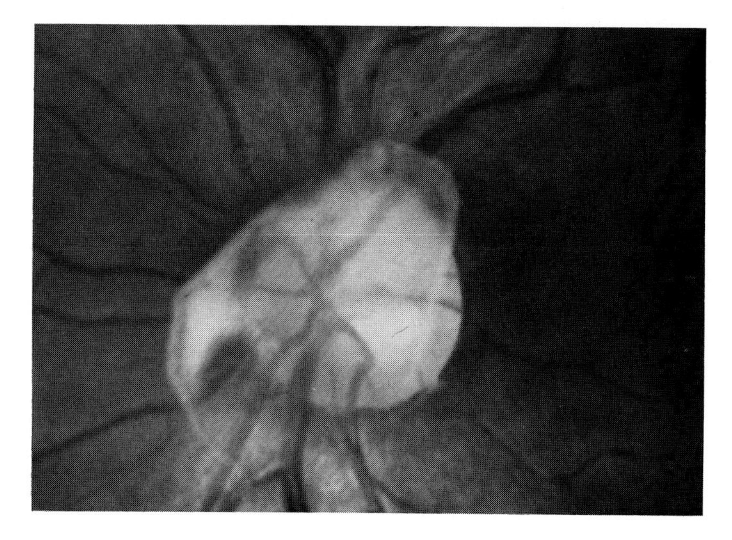

Fig. 2–9. Bergmeister's papilla in the form of prepapillary membrane.

sodermal cells at the posterior pole of the lens, thus forming the posterior tunica vasculosa lentis (8- to 9-mm stage).

This posterior segment extends laterally and anteriorly and eventually forms the anterior tunica vasculosa lentis (17- to 25-mm stage). During the 40- to 60-mm stage, elaborate proliferation of this system occurs, filling the entire cavity of the optic cup so that the vasa hyaloidea propria is formed. The vascular network provides vascularization for the entire interior of the developing eye.

After reaching a remarkably high stage of differentiation, the fetal hyaloid vascular system begins to atrophy (60 mm). By 8½ months, the atrophy of these vessels is almost complete (Fig. 2–10). During the seventh month, Bergmeister's ectodermal papilla also begins to atrophy, ultimately leaving a physiologic cup of the disc.

Agatston, S.A.: Congenital cyst of the optic nerve. Am. J. Ophthalmol., *27*:278–279, 1944.

Bech, K., and Jensen, O.A.: Racemose haemangioma

of the retina. Acta Ophthalmol. (Copenh.), *36*:769, 1958.

Bisland, T.: Vascular loops in the vitreous. Arch. Ophthalmol., *49*:514-529, 1953.

Blodi, F.C.: Preretinal glial nodules in persistence and hyperplasia of primary vitreous. Arch. Ophthalmol., *87*:531–534, 1972.

De Beck, D.: Persistent remains of fetal hyaloid system. Monogr. Am. Ophthalmol., *3*:1–78, 1890.

Degenhart, W., et al.: Congenital vascular anomalies of the optic nerve head. Trans. Pac. Acad. Ophthalmol. Otolaryngol., *34*:152–157, 1981.

Delaney, W.V., Jr.: Prepapillary hemorrhage and persistent hyaloid artery. Am. J. Ophthalmol., *90*:419–421, 1980.

Duke-Elder, S.: Anomalies of the fetal vascular system. *In* System of Ophthalmology. Vol. 3. St. Louis, C.V. Mosby, 1963. Pt. 2.

Easterbrook, M., and Sloan, A.: Anomalies of the human hyaloid system. Can. J. Ophthalmol., *13*:283–286, 1978.

Kingham, J.D.: Persistent hyaloid with retinal detachment. Ophthalmic Surg., *8*:82–87, 1977.

Hamming, N.A., et al.: Ultrastructure of the hyaloid vasculature in primates. Invest. Ophthalmol. Vis. Sci., *16*:408–415, 1977.

Hilgartner, H.L.: Benign cyst of the optic disc. Am.J. Ophthalmol., *23*:186–187, 1940.

Jones, E.H.: Hyaloid remnants in the eyes of premature babies. Br. J. Ophthalmol., *23*:186–187, 1940.

Levitt, J.M., and Lloyd, R.I.: Congenital pre-papillary cyst containing a moving vascular loop. Am. J. Ophthalmol., *22*:760–764, 1939.

Mann, I.: Developmental Abnormalities of the Eye. Cambridge, University Press, 1937.

Primrose, J.: Triple branching of retinal blood vessel. Br. J. Ophthalmol., *44*:246–247, 1960.

Pruett, R.G.: The pleomorphism and complications of posterior hyperplastic primary vitreous. Am. J. Ophthalmol., *80*:625–629, 1975.

Renz, B.E., and Vygantas, C.M.: Hyaloid vascular remnants in human neonates. Ann. Ophthalmol., *9*:179–184, 1977.

Snodgrass, M.B.: Anomalous development of retinal vessels. Br. J. Ophthalmol., *40*:754–755, 1956.

Tower, P.: Congenital prepapillary cyst. Arch. Ophthalmol., *48*:433–435, 1952.

Tyner, G.S.: Ophthalmoscopic findings in normal premature infants. Arch. Ophthalmol., *45*:627–629, 1951.

Walsh, F.B., and Hoyt, W.E.: Clinical Neuro-Ophthalmology. 3rd Ed., Vols. 1–3. Baltimore, Williams & Wilkins, 1969.

Cups, Craters, Colobomas, and Crescents

PHYSIOLOGIC CUPPING

The degree of physiologic cupping is determined conjointly by the size of the scleral ring and by the degree of atrophy or dissolution of the tissues of Bergmeister's papilla. A small depression that occupies 20 to 30% of the disc's central area is the rule, with the stippled lamina cribrosa visible in its base 0.5 to 1.0 mm below the plane of the nerve. There may be a small scleral ring with essentially no central cup or a larger scleral ring with broad cupping.

Variations in Cupping

Occasionally, the "physiologic" cup may be greatly exaggerated in width and depth; it may occupy 80% or more of the disc area and extend posteriorly 2 to 3 mm. This characteristic may represent a central intrapapillary coloboma or "crater" of the optic nerve head and may resemble the pathologic cupping of glaucoma (Fig. 2–11). In the congenital variant, the cup is confined within the boundaries of the disc itself and does not involve the peripheral rim of normal nerve tissue, except rarely on the temporal side. In a glaucomatous cup, the entire edge of the disc is undermined; this is usually more prominent nasally.

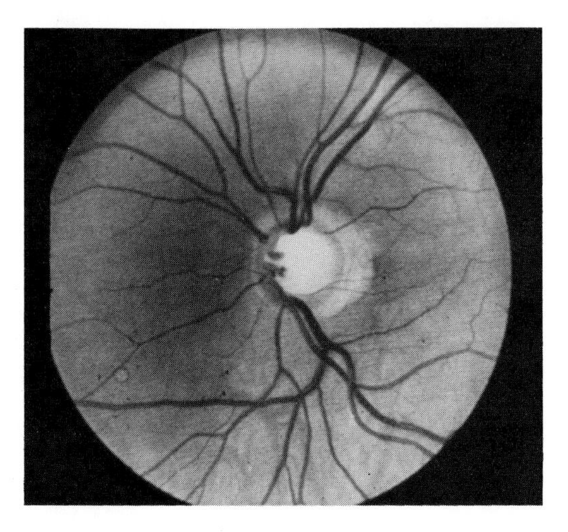

Fig. 2–11. Congenital intrapapillary coloboma, or central disc crater, which resembles the pathologic cupping of glaucoma. An associated peripapillary crescent also appears.

DISC COLOBOMAS

A coloboma of the optic disc represents a defect of the primitive optic papilla and may involve part or all of the nerve head's substance (Fig. 2–12). Visual acuity may be either severely depressed or entirely normal. Colobomas that involve the inferior portion of the optic disc are often associated with other colobomas below the disc and occasionally with similar midline defects of the lens and iris (Fig. 2–13).

Peripapillary Staphyloma

This is a related condition and probably reflects a defective closure of the proximal fetal fissure or developmental weakness of the posterior sclera. The optic disc is often seen to lie at the bottom of a posterior scleral bulging in the posterior pole. The disc may be normal in appearance or grossly malformed; it is surrounded by stretched choroid that exposes bare sclera (Fig. 2–14). Visual function is usally poor. A type of contractile staphyloma with arrhythmic movements of its wall also have been described, although rarely. This staphyloma is probably caused by an atavistic retractor bulbi muscle.

Combined Coloboma-Staphyloma

This condition, which has been described as the "morning glory" anomaly, consists of an optic disc coloboma combined with a peripapillary staphyloma, and an abnormal pattern of branching and exiting of the central retinal vessels from the nerve head (Fig. 2–15). It is usually associated with axial myopia and amblyopia and is most often unilateral.

Congenital Pit or Hole

Congenital pits (holes) of the optic nerve usually occur most in the inferior temporal

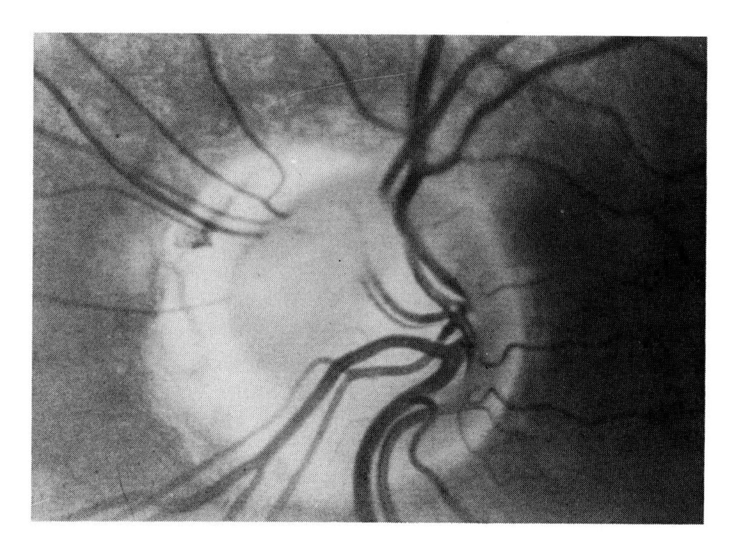

Fig. 2–12. Coloboma of the optic nerve with temporal crescent.

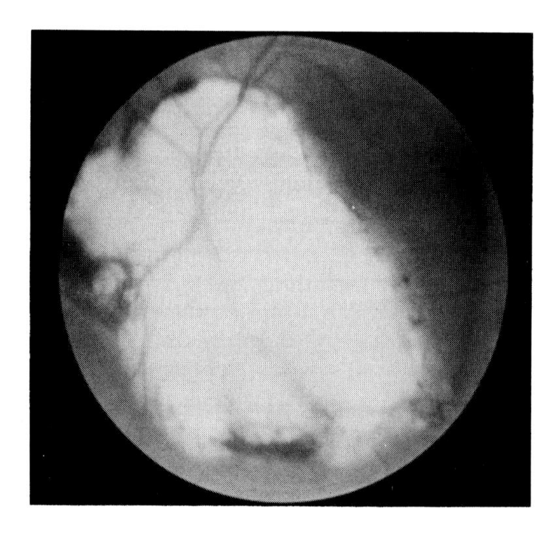

Fig. 2–13. Coloboma of the disc and inferior choroid.

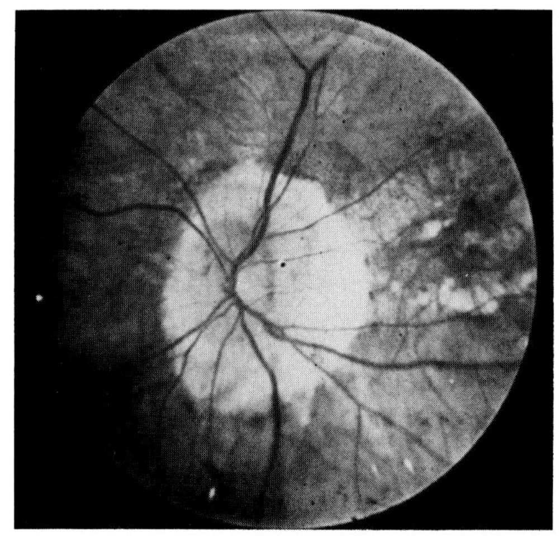

Fig. 2–14. Peripapillary staphyloma with "stretch marks" of the surrounding choroid.

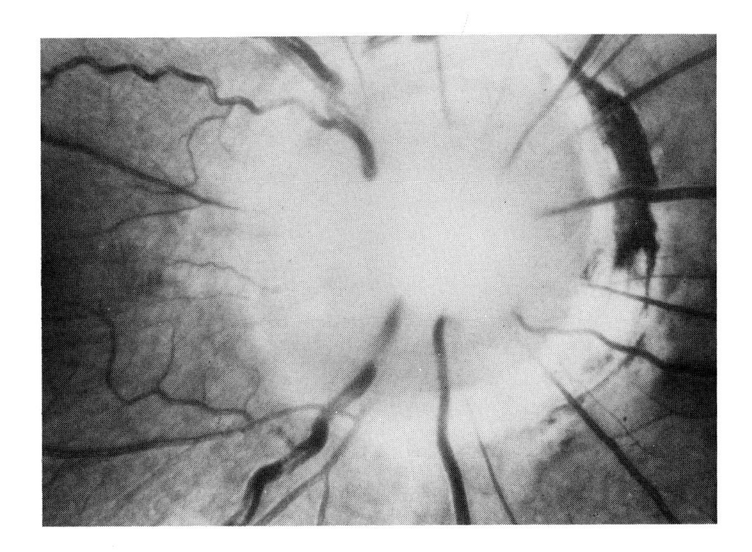

Fig. 2–15. "Morning glory" disc. Coloboma of the disc with staphyloma of the posterior pole and anomalous branching of the central retinal vessels.

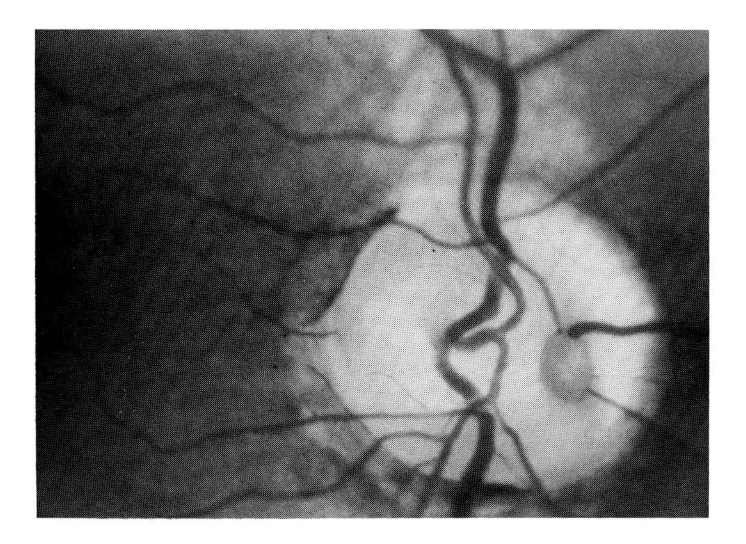

Fig. 2–16. Optic disc pit in the inferior temporal quadrant.

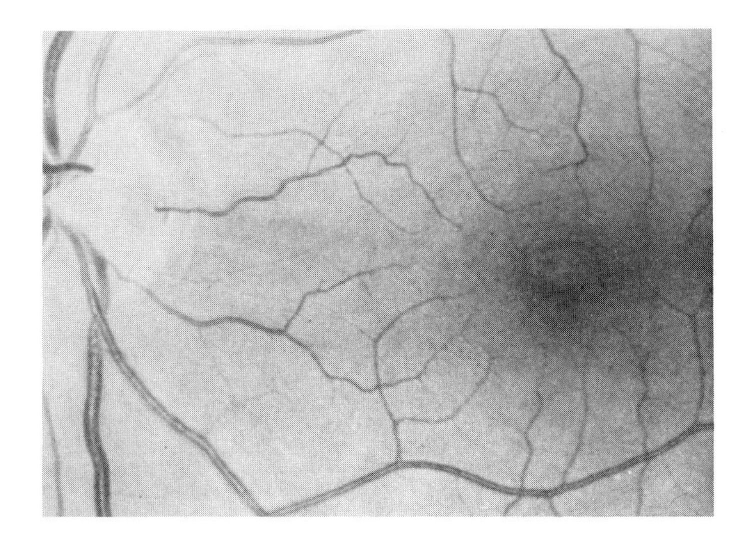

Fig. 2–17. Temporal optic disc pit with subretinal fluid beneath the macula.

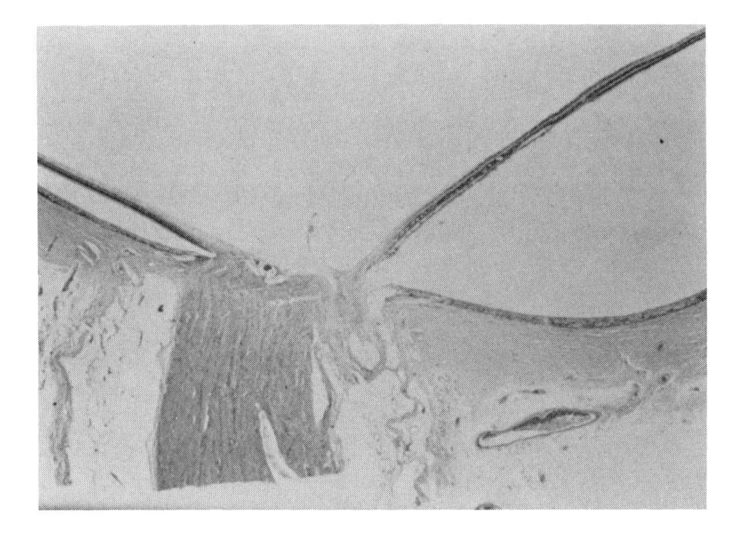

Fig. 2–18. Microscopic appearance of optic disc pit demonstrating the intramedullary defect extending to the meningeal-vaginal space.

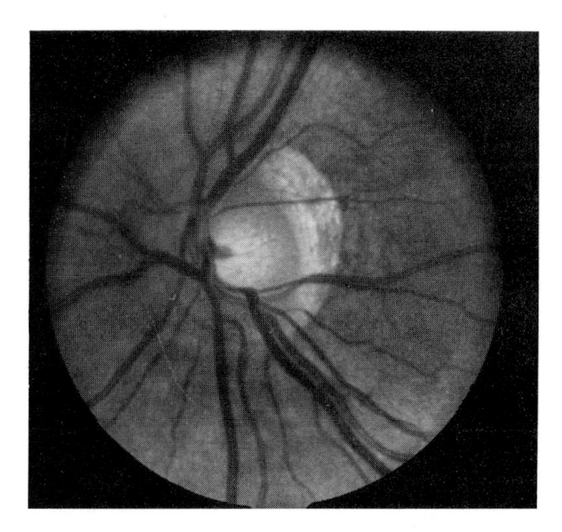

Fig. 2–19. Temporal myopic crescent.

portion of the optic disc and often contain pigmented glial tissue, which provides the disc with a grayish appearance (Fig. 2–16). Most authors have regarded these pits as atypical "minimal" colobomas or as a meningocele of the disc. Congenital pits are usually unilateral and single, but multiple pits have been described.

Almost all types of field defects have been associated with this type of disc anomaly, emanating most commonly from the blind spot. Serous detachment of the macula as a complication of congenital pits is frequent (Fig. 2–17). The source of fluid responsible for macular detachment has not been firmly identified. Fluid vitreous may leak directly through the hole and into the subretinal space. This complication usually occurs during the second and third decades of life.

Optic nerve pits have also been described as occurring concomitantly with tilted hypoplastic discs, discs with colobomas, or intermittent subretinal fluid accumulation.

The pit is formed by rudimentary retinal tissue, glial elements, nerve fibers, and pigment epithelium. These dip down into the intramedullary tissue toward the vaginal sheath, where they come into contact with the sheath's inner (neural) aspect. In the region of the pit, the lamina cribrosa is defective, and the retinal nerve fibers skirt its margin as they exit through the optic nerve (Fig. 2–18).

CRESCENTS AND CONUS

Congenital crescents are common and often are associated with myopia, optic nerve hypoplasia, and the tilted nerve.

The myopic crescent is almost always on the temporal side of the optic disc (Fig. 2–19). In such a myopic eye, the scleral canal is oblique, and the crescent seen is the inner

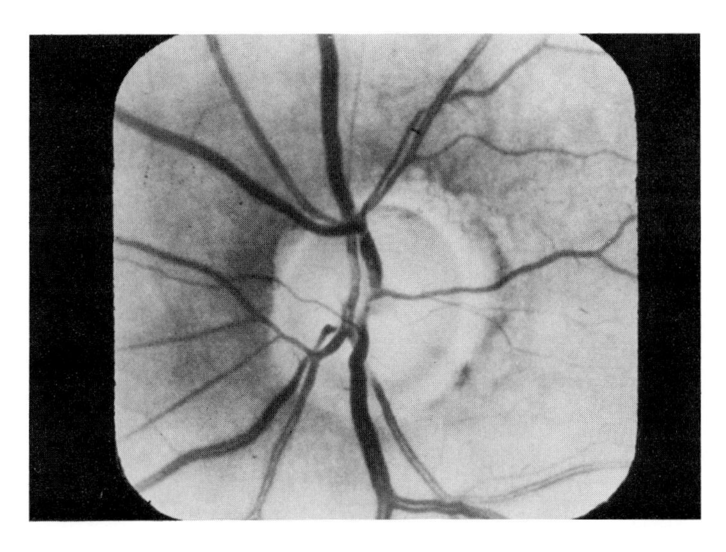

Fig. 2–20. Peripapillary crescent in an eye with high axial myopia with choroid displacement from disc margin.

surface of the canal. In other cases, the crescent is secondary to stretching of the globe and displacement of the choroid away from the margin of the disc and may completely surround the disc as a peripapillary crescent (Fig. 2–20).

True congenital crescents usually occur inferiorly and often are associated with the tilt-conus-ectasia form of optic nerve hypoplasia, with variable visual function, and with upper and more temporal field defects. It is important to distinguish these visual field defects from the true bitemporal field defects associated with chiasmal lesions, which never cross the vertical meridian to extend into the upper nasal quadrant.

RELATED TERMS

(1) Congenital cavitation
(2) Conus
(3) "Morning glory" disc
(4) Myopic crescent
(5) Optic disc dysversion
(6) Optic nerve coloboma
(7) Optic nerve hypoplasia/dysplasia
(8) Optic nerve pit
(9) Peripapillary coloboma
(10) Posterior staphyloma
(11) Situs inversus opticus
(12) Temporal crescent
(13) Tilt-conus-ectasia syndrome

Brink, J.K., and Larsen, F.E.: Pseudodoubling of the optic disc. A fluorescein angiographic study of a case with coloboma. Acta Ophthalmol. (Copenh.), *55*:862–870, 1977.
Brockhurst, R.J.: Optic pits and posterior retinal detachment. Trans. Am. Ophthalmol. Soc., *73*:264–291, 1975.
Brown, G.C., and Augsburger, J.J.: Congenital pits of the optic nerve head and retinochoroidal colobomas. Can. J. Ophthalmol., *15*:144–146, 1980.
Brown, G.C., Shields, J.A., and Goldberg, R.E.: Congenital pits of the optic nerve head. II. Clinical studies in humans. Ophthalmol., *87*:51–65, 1980.
Brown, G.C., et al.: Congenital pits of the optic nerve head. I. Experimental studies in collie dogs. Arch. Ophthalmol., *97*:1341–1344, 1979.
Brown, G.C., et al.: Congenital pits and serous retinal detachment. Trans. Pacif. Acad. Ophthalmol. Otolaryngol., *32*:151–154, 1979.
Carpel, E.F., and Engstrom, P.F.: The normal cup-disk ratio. Am. J. Ophthalmol., *91*:588–597, 1981.
Chang, M.: Pits and crater-like holes of the optic disc. Ophthalmol. Sem., *1*:21–61, 1976.
Corbett, J.J., et al.: Cavitary development defects of the optic disc. Visual loss associated with optic pits and colobomas. Arch. Neurol., *37*:210–213, 1980.
Duke-Elder, S.: Congenital crescent (conus). *In* System of Ophthalmology. Vol. 3. St. Louis, C.V. Mosby, 1963, Pt. 2.
Font, R.L., and Zimmerman, L.E.: Intraocular scleral smooth muscle in coloboma of the optic disc. Am. J. Ophthalmol., *72*:452, 1971.
Godel, V., and Regenbogen, L.: Unilateral retinitis pigmentosa and pit of the optic disc. Arch. Ophthalmol., *94*:1417–1418, 1976.
Goldberg, R.E.: Optic nerve pit and associated coloboma with serous detachment. Arch . Ophthalmol., *97*:160–161, 1974.
Grimson, B.S., Mann, J.D., and Pantell, J.P.: Optic nerve pit during papilledema. Arch. Ophthalmol., *100*:99–100, 1982.
Hayreh, S.S., and Cullen, J.F.: Atypical minimal peripapillary choroidal colobomata. Br. J. Ophthalmol., *56*:86–96, 1972.
Kayazawa, F.: A case of an optic pit. Ann. Ophthalmol., *13*:865–868, 1981.
Kindler, P.: Morning glory syndrome: Unusual congenital optic disc anomaly. Am. J. Ophthalmol., *69*:376–384, 1970.
Magli, A., Ambrosio, G., and Francois, J.: Ocular coloboma with congenital heart disease in the absence of chromosomal abnormalities. Ophthalmologica, *181*:195–202, 1980.
Mann, I.: Developmental Abnormalities of the Eye. Cambridge, University Press, 1937.
Mustonen, E., and Varonen, T.: Congenital pit of the optic nerve head associated with serous detachment of the macula. Acta Ophthalmol. (Copenh.), *50*:689–698, 1972.
Pagon, R.A.: Ocular coloboma. Surv. Ophthalmol., *25*:223–236, 1981.
Pfaffenbach, D.D., and Walsh, F.B.: Central pit of the optic disk. Am. J. Ophthalmol., *73*:102–106, 1972.
Rack, J.H., and Wright, G.F.: Coloboma of the optic nerve entrance. Br. J. Ophthalmol., *50*:705–709, 1966.
Rosen, E.: Crater-like holes in optic disc. Br. J. Ophthalmol., *32*:465–478, 1948.
Walsh, F.B., and Hoyt, W.F.: Normal and abnormal cupping. *In* Clinical Neuro-Ophthalmology. 3rd Ed. Vol. 1. Baltimore, Williams & Wilkins, 1969.
Wise, J.B., MacLean, A.L., and Gass, J.D.: Contractile peripapillary staphyloma. Arch. Ophthalmol., *75*:626–630, 1965.
Wright, P.: Meningocele of the optic disc. Br. J. Ophthalmol., *44*:570–571, 1960.

Elevated Disc Anomalies

PSEUDOPAPILLEDEMA

"Pseudopapilledema" is a nonspecific term used to describe any anomalous elevation of the optic disc that is superficially similar in appearance to papilledema. Visual acuity is not usually affected, nor is there evidence of venous stasis retinopathy. The patient is most often asymptomatic and free of other signs and symptoms of increased intracranial pressure.

Pseudoneuritis

Another closely related term, "pseudoneuritis," is applied to elevated disc anomalies with anomalous vessels, glial remnants of the disc, and mildly defective vision, that is often caused by uncorrected hyperopia and astigmatism. Figure 2–21 shows the clinical appearance of the "pseudoneuritis" type of pseudopapilledema, with anomalous vessels and hyperopia.

Drusen

Drusen buried beneath the surface of the optic nerve and anterior to the lamina cribrosa account for approximately 80% of the diagnostically troublesome elevated disc anomalies of childhood (Fig. 2–22). Rarely in infancy, but usually by the second or third decade, the drusen may become visible on the surface of the disc; occasionally, however, they remain "buried" for the patient's lifetime. Field defects (principally, nerve fiber bundle defects) may appear in young adult life and then slowly progress. An occasional patient will demonstrate acute or subacute visual loss secondary to hemorrhage, ischemia, or edema around the drusen (Fig. 2–23).

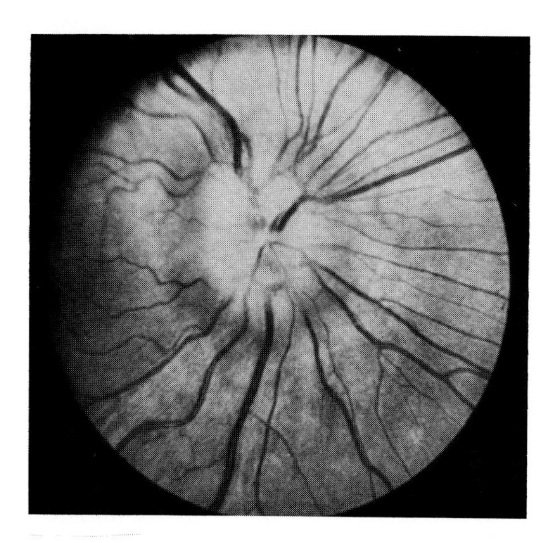

Fig. 2–21. Pseudopapilledema. "Pseudoneuritis" with anomalous vessels and hyperopia.

An irregular dominant hereditary pattern is common, and examination of the patient's parents may provide helpful clues. Drusen seem to occur almost exclusively in Caucasians, rarely in Orientals or blacks.

Disc elevation from intrapapillary drusen is seldom uniform; it appears more often as irregular elevation confined to the disc without involving the retina and forming sharp and steep disc margins. Venous engorgement usually is not evident as in true papilledema, but it may occur rarely as an associated anomaly.

Superficial drusen are round, globular, reflectile bodies of hyaline material and calcium and are easily detectable with the ophthalmoscope (Fig. 2–24). They also may demonstrate autofluorescence with the cobalt blue filter.

Recent technologic developments in the areas of ultrasonography and computed tomography have helped significantly to identify buried drusen as a cause of elevated optic nerve anomalies (Fig. 2–25).

Current theory holds that drusen arise as a result of reduced axoplasmic flow, which leads to neuron degeneration and release of mitochondria, with subsequent gliosis, hyalinization, and calcification (Fig. 2–26). These are not the same as the calcified astrocytomas seen in tuberous sclerosis.

Drusen have also been recognized to occur in chronic atrophic optic nerves. I have seen three unilateral cases of drusen formation

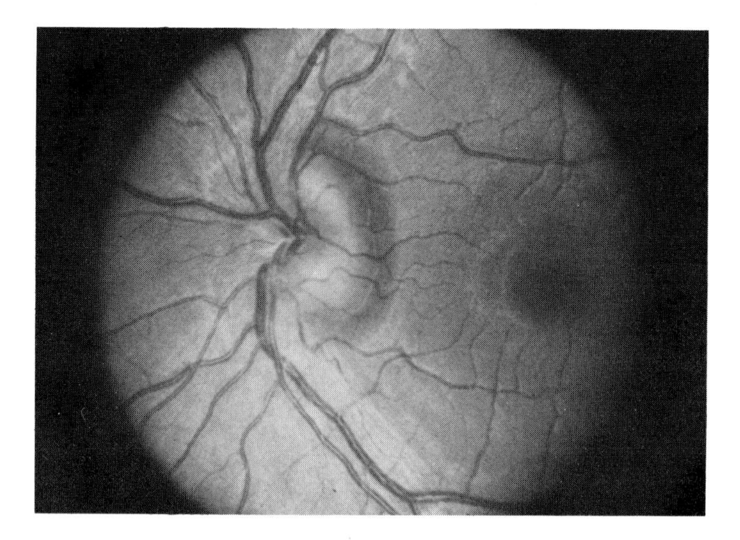

Fig. 2–22. Pseudopapilledema. Buried drusen of the optic nerve.

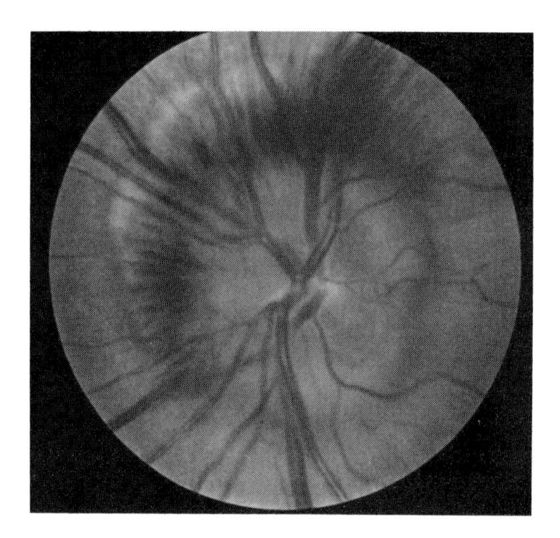

Fig. 2–23. Drusen with peripapillary hemorrhage and distortion of vision.

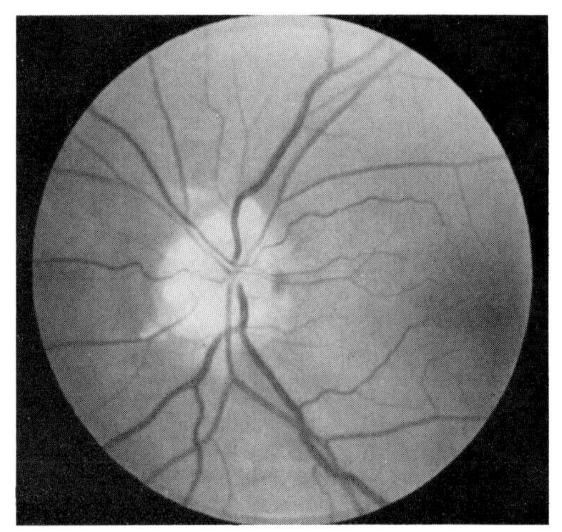

Fig. 2–24. Superficial drusen causing irregular elevation of the nerve head.

in such atrophic nerves, secondary to intraorbital meningiomas. Unilateral drusen in the presence of acquired optic atrophy may therefore herald more than just an interesting abnormality.

Hyaline bodies, or drusen, of the optic papilla are known to occur in the following circumstances:

(1) Sporadically
(2) With a family history
(3) With retinitis pigmentosa
(4) With optic atrophy
(5) With pseudoxanthoma elasticum
(6) With hypermetropia
(7) Following acute or chronic disc swelling

Hyperopic Pseudopapilledema

Perhaps the rarest type of pseudopapilledema is that type of disc elevation associated with moderate to severe degrees of hyperopia. The association of disc elevation

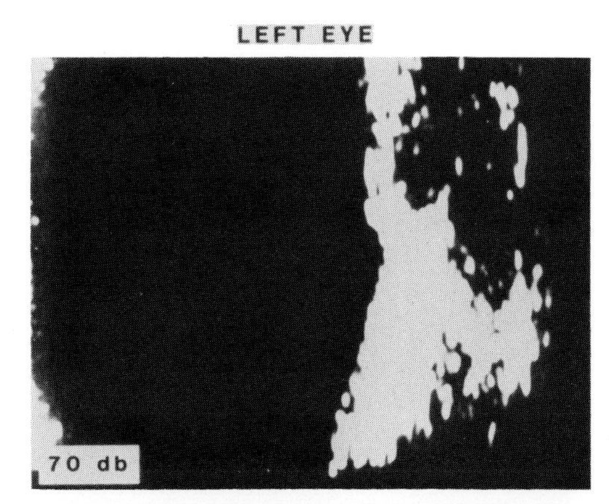

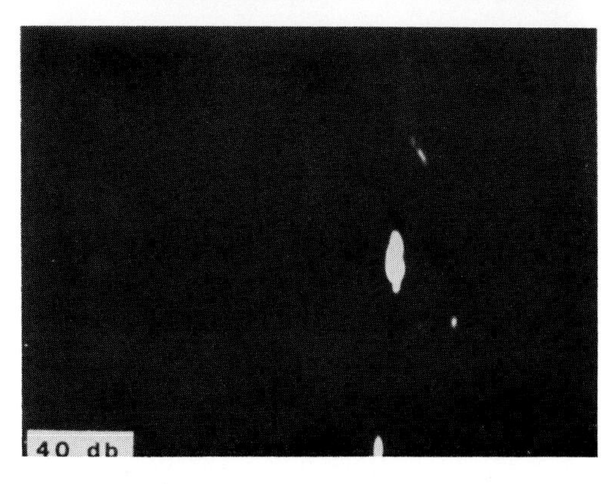

Fig. 2–25. Ultrasonogram of buried drusen in the left eye. The right eye's ultrasonogram was similar. When the decibel sensitivity is decreased from 70 dB to 40 dB, the drusen are revealed since the background optic nerve tissue is deleted.

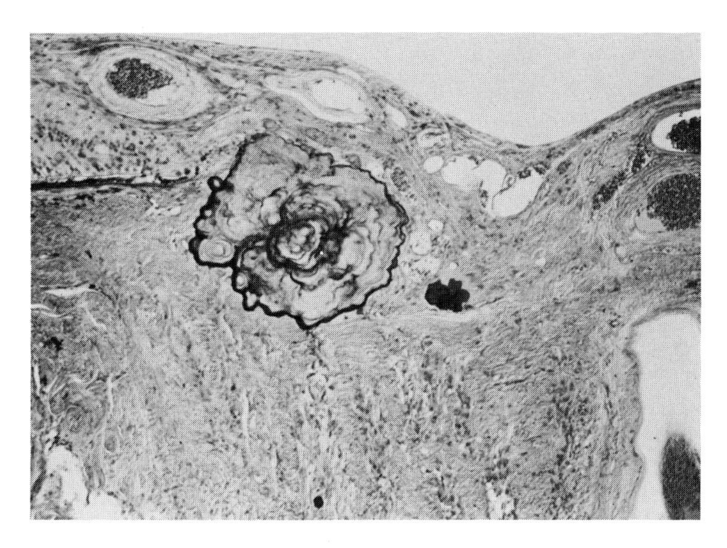

Fig. 2–26. Histopathologic specimen of optic nerve drusen, showing calcified drusen anterior to the lamina cribrosa.

with hyperopia, general good health, and normal-size blind spots suggests an anomalous disc. Such a disc anomaly is stationary and is not usually associated with venous engorgement or tortuosity. The elevation may be caused by a small scleral canal with crowding of the nerve fibers or, perhaps, by a congenital gliosis of the nerve as described with "pseudoneuritis."

Behrens, M.M.: Congenital blurred or elevated disc anomalies. *In* Clinical Ophthalmology. Vol. 2. Edited by T.D. Duane. Philadelphia, Harper & Row, 1981.

Bishara, S., and Feinsod, M.: Visual evoked response as an aid in diagnosing optic nerve head drusen: Case report. J. Pediatr. Ophthalmol. Strabismus, *17*:396–398, 1980.

Calhoun, E.P.: Pseudo optic neuritis. Ophthalmic Res., *23*:226–229, 1914.

Chambers, J.W., and Walsh, F.B.: Hyaline bodies in the optic discs: Importance in neurological diagnosis. Brain, *74*:95–108, 1951.

Chamlin, M., and Davidoff, L.M.: Drusen of optic nerve simulating papilledema. J. Neurosurg., *7*:70–78, 1950.

Erkkila, H.: Pseudopapilledema. Arch. Ophthalmol., *97*:1543, 1979.

Erkkila, H.: Clinical appearance of optic disc drusen in childhood. Albrecht Von Graefes Arch. Klin. Exp. Ophthalmol., *193*:1–18, 1975.

Glaser, J.S.: Anomalous disc elevation and hyaline bodies: Pseudopapilledema. *In* Clinical Ophthalmology. Vol. 2. Edited by T.D. Duane. Philadelphia, Harper & Row, 1981.

Gutteridge, I.F.: Optic nerve drusen and pseudopapilledema. Am. J. Optom. Physiol. Opt., *58*:671–676, 1981.

Harris, M.J., Fine, S.L., and Owens, S.L.: Hemorrhagic complications of optic nerve drusen. Am. J. Ophthalmol., *92*:70–76, 1981.

Hepler, R.S.: Pseudopapilledema (letter). Arch Ophthalmol., *97*:1543, 1979.

Hoyt, W.F., and Pont, M.E.: Pseudopapilledema: Anomalous elevation of optic disk. Pitfalls in diagnosis and management. J.A.M.A., *181*:191–196, 1962.

Kelley, J.S.: Autofluorescence of drusen of the optic nerve head. Arch. Ophthalmol., *92*:263–264, 1974.

Lambert, R., and McDannald, C.E.: Hereditary high myopia. Am. J. Ophthalmol., *14*:46–48, 1931.

Levine, R.A., et al.: Principles and Practice of Ophthalmology. Vol. 3. Edited by Gholam A. Peyman et al. Philadelphia, W.B. Saunders, 1980, pp. 2130–2132.

Lorentzen, S.E.: Drusen of the optic disk: A clinical and genetic study. Acta Ophthalmol. [Suppl.] (Copenh.), *90*:1–181, 1966.

Okun, E.: Chronic papilledema simulating hyaline bodies of the optic disc. Am. J. Ophthalmol., *53*:922–927, 1963.

Primrose, J.A.E.: Pseudopapilledema and pseudoneuritis. Proc. R. Soc. Med., *58*:116–119, 1965.

Rosenberg, M.A., Savino, P.J., and Glaser, J.S.: A clinical analysis of pseudopapilledema. I. Population, laterality, acuity, refractive error, ophthalmoscopic characteristics, and coincident disease. Arch. Ophthalmol., *97*:65–70, 1979.

Rubenstein, K., and Ali, M.: Retinal complications of optic disc drusen. Br. J. Ophthalmol., *66*:83–95, 1982.

Sacks, J.G., and Choroinokos, E.: Drusen of the optic disc. *In* Neuro-Ophthalmology Update. Edited by J.L. Smith. New York, Masson, 1977.

Saraus, H., et al.: Hemorrhages of drusen with pseudopapilledema. J. Fr. Ophtalmol., *3*:647–651, 1980.

Savino, P.J., and Glaser, J.S.: Pseudopapilledema versus papilledema. Internatl. Ophthalmol. Clin., *17*:115–137, 1977.

Savino, P.J., Glaser, J.S., and Rosenberg, M.A.: A clinical analysis of pseudopapilledema. II. Visual field defects. Arch. Ophthalmol., *97*:71–75, 1979.

Spencer, W.H.: Drusen of the optic disc and aberrant axoplasmic transport (XXIV Edward Jackson Memorial Lecture). Am. J. Ophthalmol., *85*:1, 1978.

Stevens, R.A., and Newman, N.M.: Abnormal visual-evoked potentials from eyes with optic nerve head drusen. Am. J. Ophthalmol., *92*:857–862, 1981.

Stiefel, J.W., and Smith, J.L.: Hyaline bodies of the optic nerve and intracranial tumor. Arch. Ophthalmol., *65*:814–816, 1961.

Tso, M.O.: Pathology and pathogenesis of drusen of the optic nerve head. Ophthalmol., *88*:1066–1079, 1981.

Wirtschafter, J.D.: Papilledema: Changing concepts and continuing challenges. *In* Neuro-Ophthalmology. Vol. 10. Edited by J.S. Glaser. St. Louis, C.V. Mosby, 1980.

Anomalies of Shape, Size, and Site of the Optic Disc

While minor degrees of variation in disc size and shape are common, gross malformations are rare. The disc is usually round or slightly oval with the vertical diameter approximately 0.2 mm longer than the horizontal diameter. Exaggerated elongation may occur in almost any axis—vertical, horizontal, or oblique—producing a "tilted disc." Rarely, the disc may even be square, triangular, polygonal, or otherwise distorted in shape.

SITUS INVERSUS

Situs inversus (dysversion or inversion) of the optic nerve head usually is seen as bilateral horizontal tilting of the nerve, with the temporal branches of the central retinal vessels emerging nasally from the disc (Fig. 2–27). The condition is probably caused by an anomalous insertion of the optic stalk into the optic vesicle or by segmental hypoplasia of the nerve.

DUPLICATION

Duplication (diastasis) of the optic disc occurs, rarely, with two discs, each provided with retinal vessels, appearing in an otherwise normal eye. The "accessory" disc may be fused with the normal disc at its margin or may be completely separate (Fig. 2–28).

HETEROTOPIA

Heterotopia of the disc is a rare condition in which the optic nerve enters the globe in an abnormal position, displaced either nasally or temporally from the sagittal. With temporal displacement, the macula also is displaced temporally from the pupillary axis, so that the eyes appear to diverge; this divergence forms a positive angle gamma. This condition has been described in association with hypertelorism.

MICROPAPILLAE

The term "micropapillae" describes a disc that functions normally and appears normal in every way except for being slightly small. The vision is usually normal or slightly reduced, and the blind spot is smaller than normal. This condition may represent a gray zone between optic nerve hypoplasia and the normal optic nerve. Grossly small optic nerves are usually hypoplastic in appear-

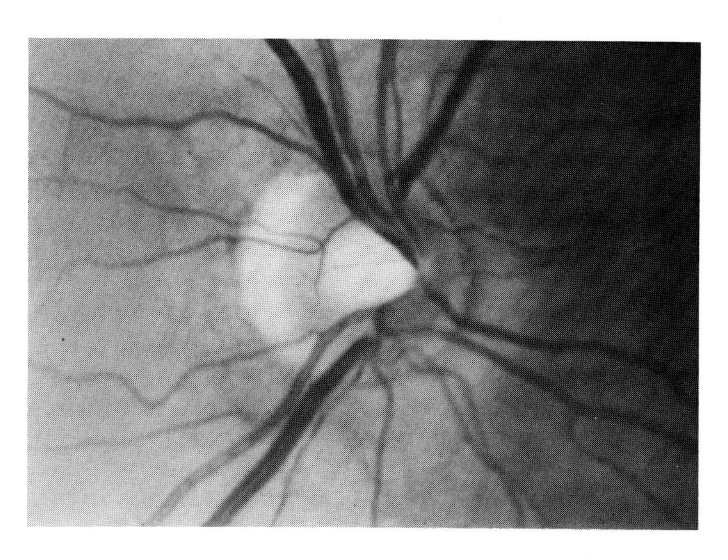

Fig. 2–27. Situs inversus of the disc (dysversion) with tilting and nasal direction of exiting vessels.

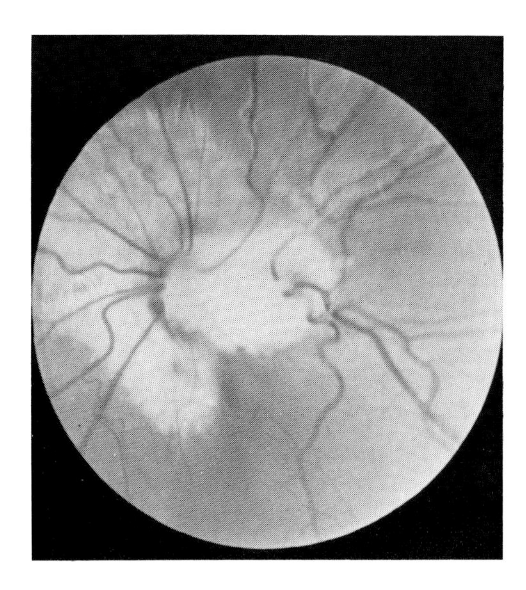

Fig. 2–28. Duplication of the disc (diastasis) with fusion of the two discs and separate retinal vessels.

ance and in visual function. Many tilted optic nerves demonstrate segmental hypoplasia with the nerve's diameter less than normal in the direction of tilt.

Total absence of the optic nerve (aplasia) is rare and is usually associated with such gross anomalies as anencephaly and anophthalmos. With total absence of the nerve, the retinal vessels are also absent and blindness is total.

MEGALOPAPILLA

Megalopapilla is a rare anomaly that demonstrates an unusually enlarged disc of up to twice the normal size. It may occur unilaterally or bilaterally with normal visual acuity and enlarged blind spots. This disorder may represent a form of congenital optic nerve gliosis.

RELATED TERMS

(1) Aplasia
(2) Diastasis
(3) Duplication
(4) Dysplasia
(5) Dysversion
(6) Heterotopia
(7) Megalopapillae
(8) Micropapillae
(9) Optic nerve dysplasia
(10) Optic nerve hypoplasia
(11) Situs inversus

Acers, T.E.: Hypertelorism, heterotopia of maculas and pseudoexotropia. Am. J. Ophthalmol., *59*:494–495, 1965.

Acers, T.E.: Optic nerve hypoplasia: Septo-optic-pituitary dysplasia syndrome. Trans. Am. Ophthalmol. Soc., *79*:425–457, 1981.

Collier, M.: Les doubles papilles optiques. Bull. Soc. Ophthalmol. Fr., *70*:328–352, 1958.

Donoso, L.A., et al.: Ocular anomalies simulating double optic disc. Can. J. Ophthalmol., *16*:84–87, 1981.

Duke-Elder, S.: Congenital excavation of the optic disc. *In* System of Ophthalmology. Vol. 3. St. Louis, C.V. Mosby, 1963, Pt. 2.

Franceschetti, A., and Bock, R.H.: Megalopapilla. Am. J. Ophthalmol., *33*:227–235, 1950.

Slade, H.W., and Weekley, R.D.: Diastasis of the optic nerve. J. Neurosurg., *14*:571–574, 1957.

Streiff, B.: On megalopapilla. Klin. Monatsbl. Augenheilkd., *139*:324, 1961.

Sylvester, P.E., and Keuser, J.: The size and growth of the human optic nerve. J. Neurol. Neurosurg. Psychiatry, *24*:45–49, 1961.

Tyner, G.S.: Ophthalmoscopic findings in normal premature infants. Arch. Ophthalmol., *45*:627–629, 1951.

Chapter 3

NEOPLASMS, HAMARTOMAS, AND VASCULAR ANOMALIES

Neoplasms and Hamartomas of the Optic Nerve

Primary tumors that affect the intraocular optic nerve papilla are rare, especially as congenital lesions. Gliomas, hemangiomas, and melanocytomas are perhaps the most common and most likely represent hamartomatous hyperplasia.

GLIOMAS

Gliomas usually occur in the first decade of life and may involve the optic nerve(s) and/or chiasm. They are usually astrocytomas or spongioblastomas. Most of these tumors arise in the intraorbital portion of the optic nerve or in the intracranial nerve and chiasm. Except in rare cases, the optic disc is not affected until later, when the presenting sign is optic atrophy. Gliomas may also occur in conjunction with the neurofibromatoses of von Recklinghausen. The clinical and microscopic appearances of the astrocytomas are shown in Figure 3–1.

HEMANGIOMAS

Hemangioma of the optic disc usually appears as a well demarcated, elevated, reddish-purple mass on the disc surface. It has been described as "a cluster of worms" or as a "raspberry"- or "mulberry"-like lesion. The condition is usually unilateral and asymptomatic. There are three basic types of hemangiomas: (1) capillary, (2) cavernous, and (3) arteriovenous malformation (Figs. 3–2, 3–3, and 3–4).

MELANOCYTOMAS

Melanocytomas are benign pigmented tumors that arise from the nerve head and occur predominantly in the black races (Fig. 3–5). They demonstrate exceptional growth, loss of their benign characteristics, and even progressive visual field defects. These tumors may be discovered in the patient at a early age, but they usually escape notice because there are no associated symptoms.

COMBINED RETINAL-PIGMENT EPITHELIAL HAMARTOMAS

These partially pigmented lesions can occur anywhere in the fundus, including the peripapillary area (Fig. 3–6). They are slightly elevated lesions with grayish-brown pigmentation and are often confused with postinflammatory scars and melanomas. Epiretinal membranes that overlie the tumor may form in any age group; when these membranes occur temporal to the disc, they may cause macular traction.

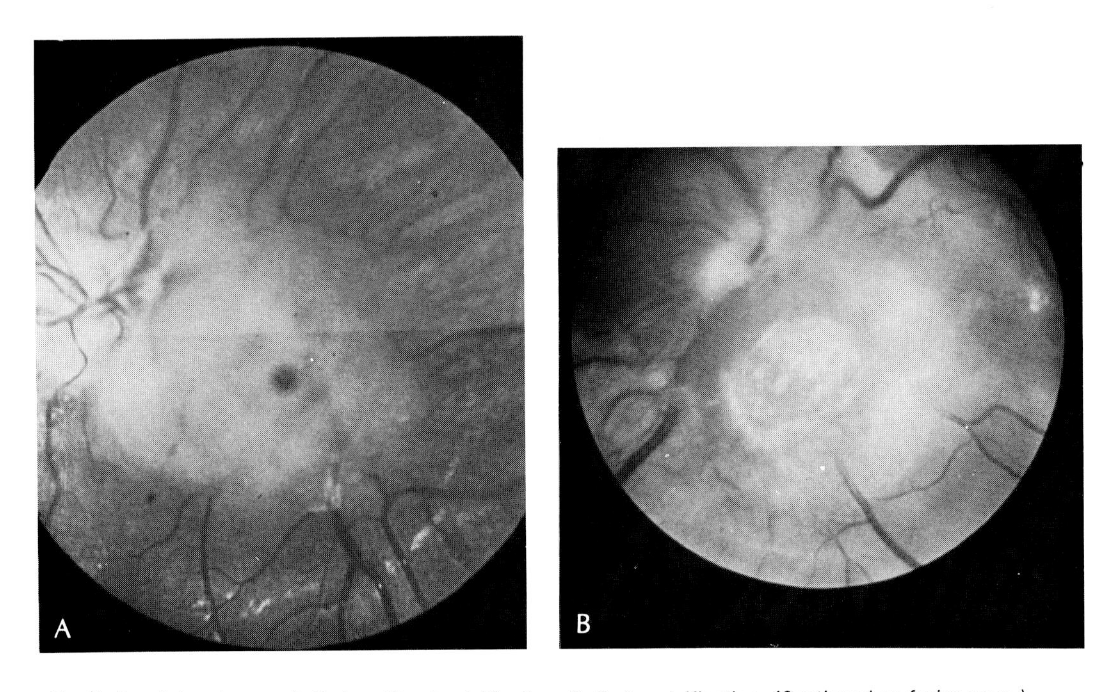

Fig. 3–1. Astrocytomas. *A.* Early, without calcification. *B.* Early calcification. *(Continued on facing page.)*

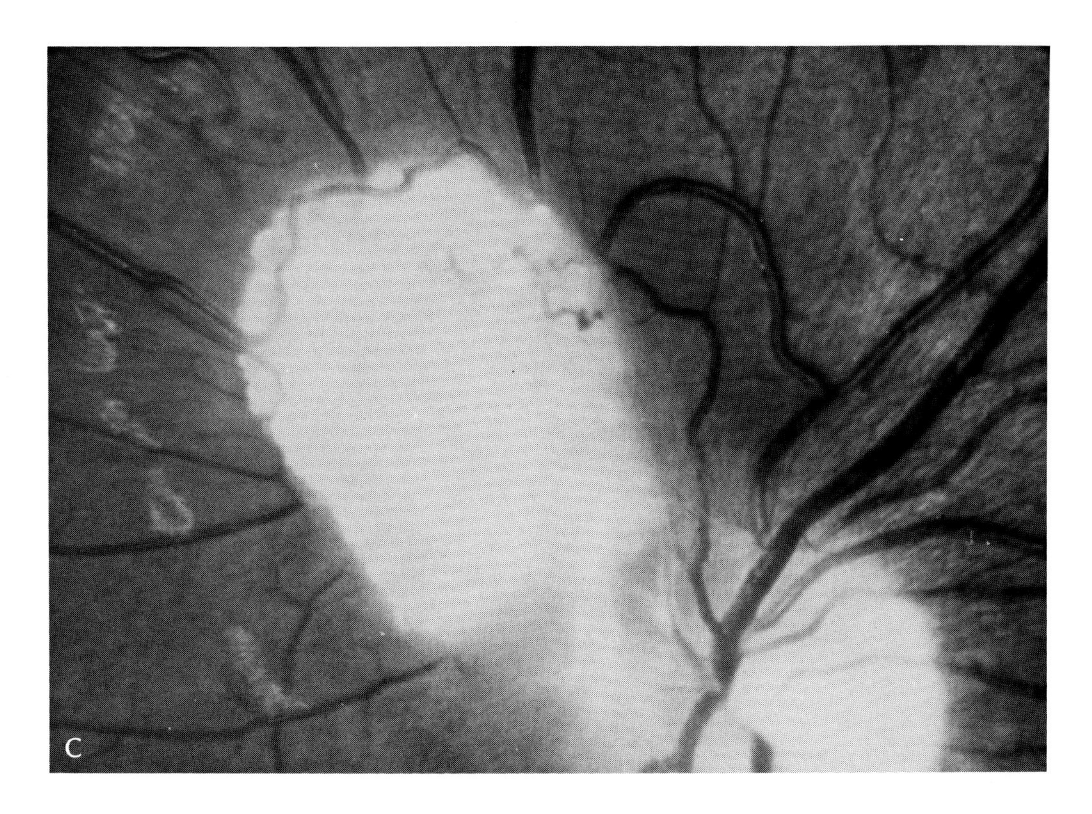

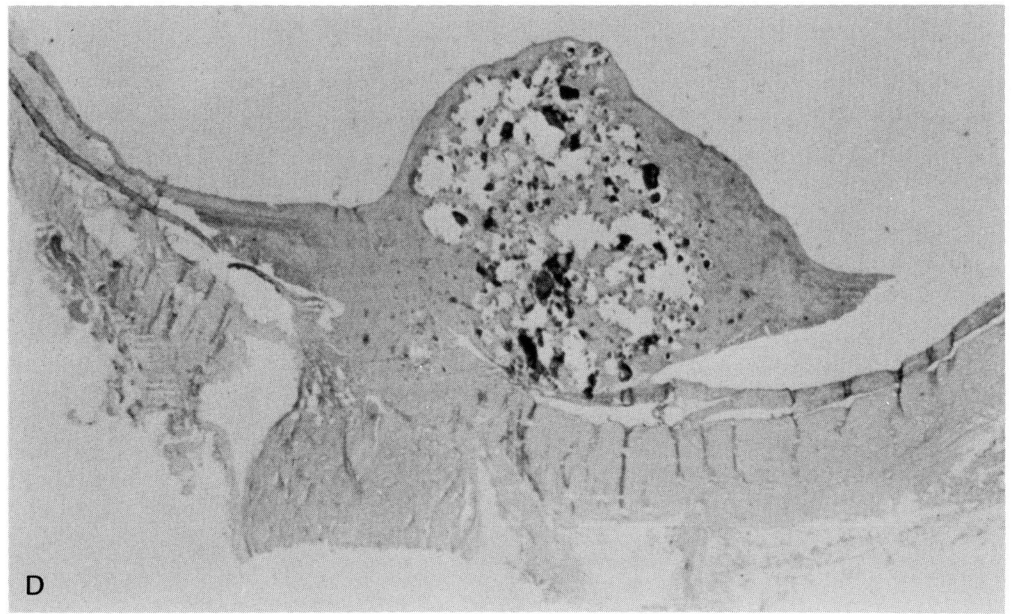

Fig. 3–1. *C.* Calcified. *D.* Histopathologic view of calcification.

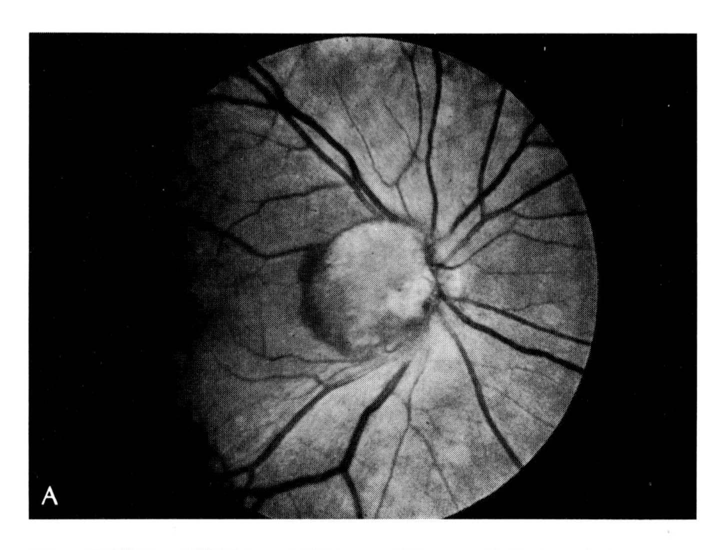

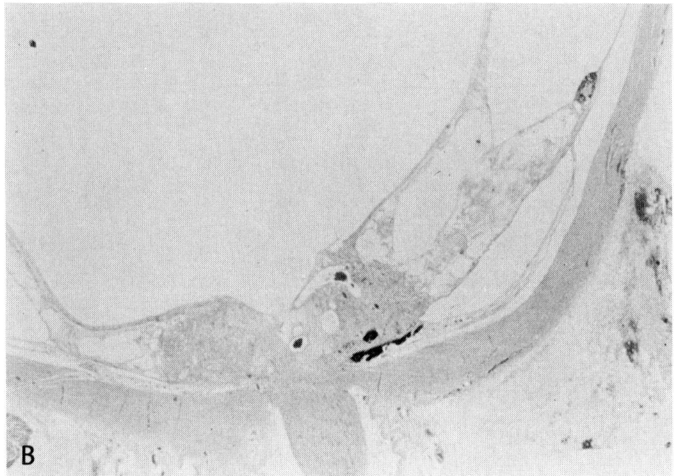

Fig. 3–2. Clinical (A) and microscopic (B) appearances of capillary hemangioma involving the nerve head and adjacent retina.

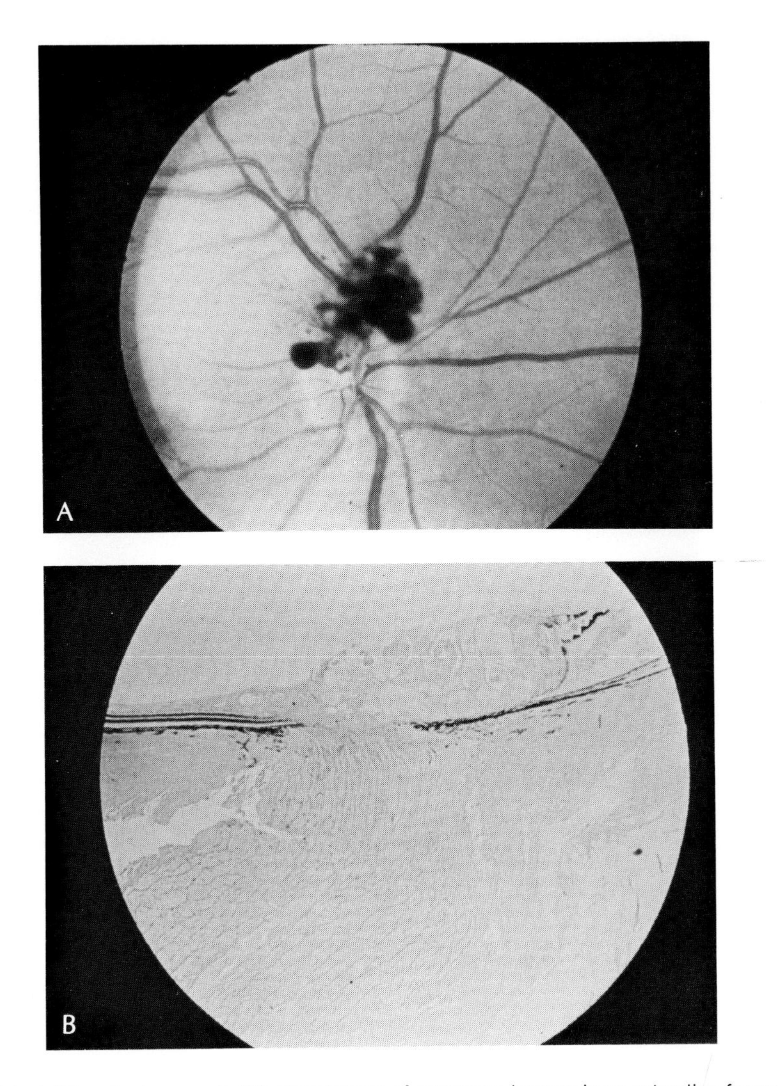

Fig. 3–3. Clinical (A) and microscopic (B) appearances of cavernous hemangioma extending forward from nerve head.

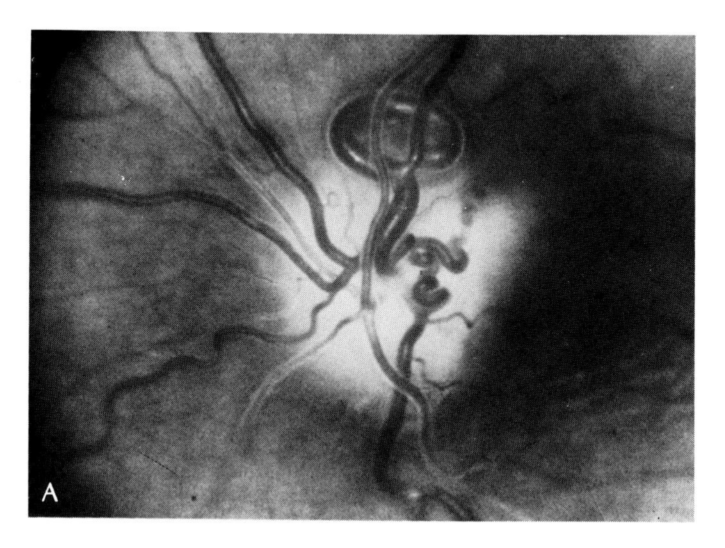

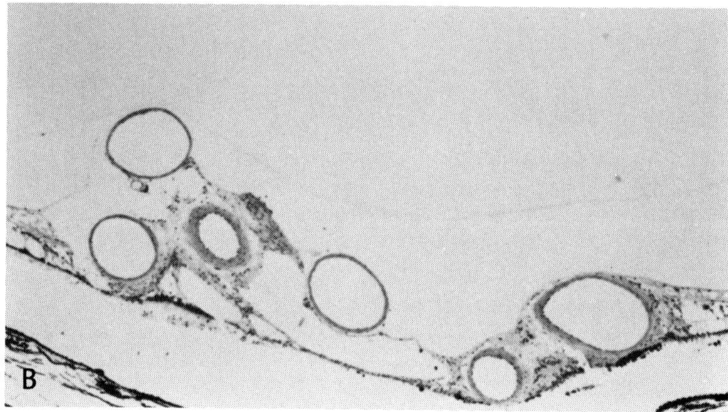

Fig. 3–4. Clinical (A) and microscopic (B) appearances of arteriovenous malformation of nerve head.

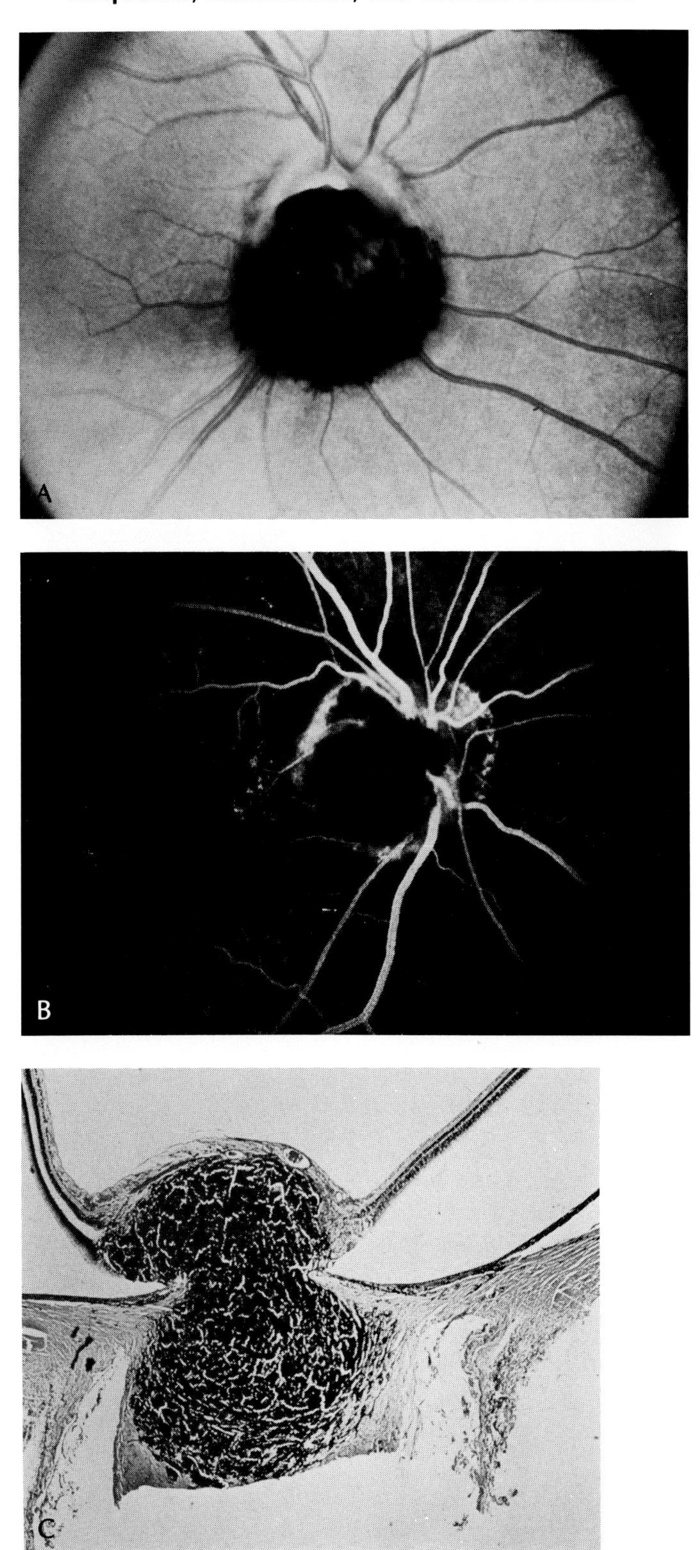

Fig. 3–5. Melanocytoma. *A.* Clinical appearance of a jet black, "furry" lesion extending from nerve head. *B.* Fluorescein angiogram demonstrating blockage of nerve head fluorescence beneath tumor. *C.* Histopathologic specimen demonstrating hypertrophy and hyperplasia of melanin-bearing cells.

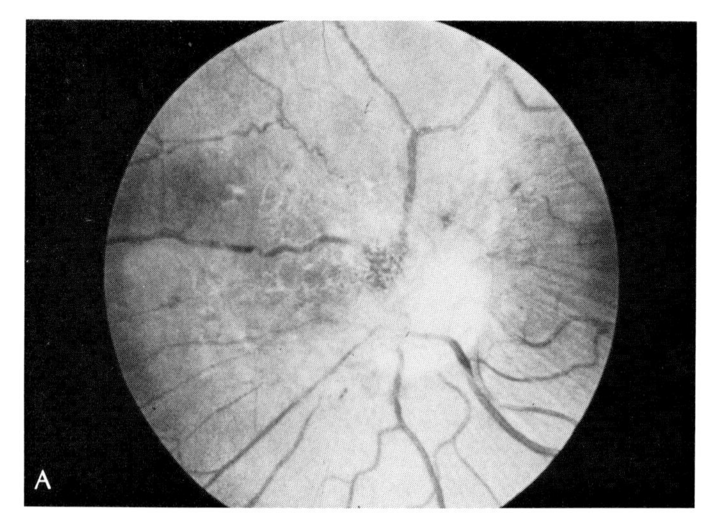

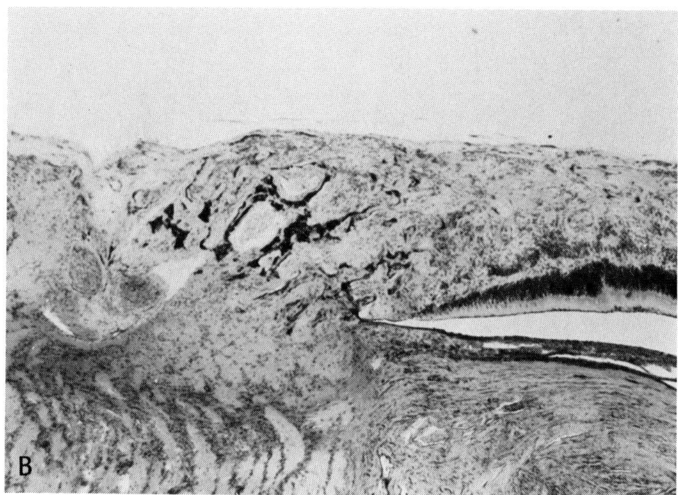

Fig. 3–6. Clinical (A) and microscopic (B) appearances of combined retinal-pigment epithelial hamartoma.

RETINOBLASTOMA

Perphaps the only true congenital and hereditary neoplasm that primarily affects the retina and extends into the optic nerve is the retinoblastoma. It is the most common intraocular tumor of childhood and approximately one third of the cases occur bilaterally. A tumor that is of considerable size and already involves the optic nerve may be present at birth, although diagnosis is usually made when the patient is two years old. Retinoblastoma is not a "primary" tumor of the optic nerve but usually spreads by direct extension through the optic nerve.

Ninety-six percent of the cases appear sporadically, but the development of the neoplasm depends on the presence of a single dominant gene. Sporadic cases occur either as a somatic mutation (90%) or as a new germinal mutation. In some cases, the tumor may be associated with a Dq-chromosome abnormality.

Retinoblastoma may occur anywhere in the retina as a solitary mass, but it usually occurs as multifocal lesions. The tumor usually appears as a white, round, nodular mass with vessels running over its irregular surface. Calcification of part of the tumor mass is common.

Retinoblastoma must be considered whenever a child presents with a retinal tumor or leukocoria (white pupillary reflex). The differential diagnosis includes:

(1) Retrolental fibroplasia: retinopathy of prematurity
(2) Coat's disease: retinal telangiectasia
(3) Persistent hyperplastic primary vitreous
(4) Retinal dysplasia
(5) Incontinentia pigmenti
(6) Embryonal medulloepithelioma: diktyoma

Other conditions in which leukocoria occurs are congenital cataracts, juvenile retinoschisis, congenital retinal detachment, medullated nerve fibers, colobomas, tuberous sclerosis, and the other phakomatoses.

The Phakomatoses: Multisystem Hamartomas

These hamartomas might be more appropriately classified as retinal anomalies if they did not also involve the optic nerve and have related CNS and systemic anomalies. Neurofibromatosis (von Recklinghausen's disease), tuberous sclerosis (Bourneville's disease), retinocerebral angiomatosis (von Hippel-Lindau disease), encephalotrigeminal angiomatosis (Sturge-Weber syndrome), and the orbital-retinal-midbrain vascular malformations are included rather loosely, along with others, among the phakomatoses (Table 3–1).

These conditions all have a congenital basis and may be apparent at birth, but as a general rule, they become more apparent in childhood. The dissimilarities of this group are sometimes more impressive than the resemblances, but historically they have been treated together.

NEUROFIBROMATOSIS

Neurofibromatosis is characterized by tumors of the skin; cutaneous pigmentation; tumors arising from the sheaths of the cranial, spinal, peripheral, and sympathetic nerves; bone abnormalities; central nervous system anomalies; and ocular anomalies. The genetic pattern is variable and has been described as sporadic, incomplete-dominant, and sex-linked recessive. In order of frequency, the ocular structures involved are lids, optic nerve, retina, iris, cornea, and conjunctiva with the lid (especially the upper lid). A form of congenital or developmental glaucoma is not rare. Gliomas of the optic nerve are definitely associated with neurofibromatosis; 15 to 20% of gliomas of the optic nerve or chiasm have other evidence of this multisystem disease. Optic atrophy is common due to primary involvement of the nerve by the glioma or secondary due to compression by adjacent neurofibroma; it may be a sequela of papilledema secondary to intracranial mass involvement. Also, there is an increased incidence of acoustic neurinomas and meningiomas in association with von Recklinghausen's disease.

TUBEROUS SCLEROSIS

Tuberous sclerosis is characterized by facial adenoma sebaceum, seizure, mental deficiency, and retinal tumors that occasionally involve the optic nerve. These tumors may appear as dull, gelatinous nodules of varying size or as multilobular, mulberry- or tapioca-like lesions. They are astrocytomas of the retina or optic nerve and may closely resemble retinoblastomas. As a general rule, the retinal vessels course through the substance of the astrocytoma and over the surface of the retinoblastoma.

RETINOCEREBRAL ANGIOMATOSIS

Angiomatosis of the retina and cerebellum also involves cystic tumors of the kidney, epididymis, and pancreas. Characteristic retinal changes include marked dilatation and engorgement of a segment of a larger vein and enlargement of the accompanying artery, which appears "beaded" and joined in a "hemangioblastoma" mass or "bee's nest." Secondary gliosis, exudation, and hemorrhages are common. These usually occur inferiorly in the periphery but may involve the posterior segments and the optic nerve. The cerebellar and/or ocular symptoms usually begin in the third or fourth decade.

ENCEPHALOTRIGEMINAL ANGIOMATOSIS

The vascular encephalotrigeminal syndrome is characterized by nevus flammeus, which is usually unilateral and is distributed through one or more branches of the trigeminal nerve (usually the ophthalmic division). Other characteristics are meningeal angiomatous involvement, seizures, intracranial calcification, and occasionally hemiplegia. With involvement of the upper lid, congenital glaucoma is not unusual. An-

TABLE 3–1. PHAKOMATOSES

SYNDROMES	FINDINGS	
	Ocular	Other
Bonnet-Dechaume-Blanc	Arteriovenous retinal angiomata, papilledema, reduced corneal sensitivity, anisocoria, strabismus	Angiomas of mesencephalon and thalamus, facial angiomata, hydrocephalus, slow speech, hemiplegia, neurologic symptoms
Bourneville*	Neurocytomas (retina and papilla)	Spongioblastomas, epilepsy, inner organs and endocrine disturbances, sebaceus adenomas, enchondromata
Hippel-von Lindau*	Retinal angiomatosis, vascular proliferation, retinal detachment, secondary glaucoma	Cerebral angiomatosis, epilepsy, psychic disturbances, dementia
Kasabach-Merritt	Conjunctival and retinal hemorrhages and hemangiomas	Large hemangiomas with splenogenic anemia, thrombocytopenic purpura
Klippel-Trenaunay-Weber	Retinal varicosity, choroidal angiomas, conjunctival telangiectasia	Vascular nevi, varicose vessels, capillary angiomas, lymphangiomas, AV aneurysms, thrombosis, dermatitis, hyperhidrosis
Louis-Bar	Conjunctival telangiectasia, pseudo-ophthalmoplegia	Cerebellar ataxia, scanning speech, mental retardation
Maffucci	Hemangioma of the lids, retinal angiomata	Dyschondroplasia with hemangiomas
Peutz-Touraine	Brown speckled dots on eye lid margins and on the conjunctiva	Brown dot-like skin pigmentations, polyposis gastrointestinal tract, hypochromic anemia, bronchial adenosis
von Recklinghausen*	Neurofibromas and neurocytomas of lid, iris, papilla, retina; hydrophthalmos	Neurofibromas inner organs and cerebral nerves, endocrine disturbances, skin neurofibromas, cafe-au-lait spots, cranial vein anomalies
Sturge-Weber*	Chorioretinal angiomata vascular proliferation, glioma, retinal detachment, conjunctival telangiectases, glaucoma	"Portwine" nevus, cerebral angiomata, hemiparesis, hemitrophy, mental retardation, epilepsy
Ward	Congenital cataracts, hypertelorism, congenital blindness, corneal opacities	Skull deformities, basal cell nevi including eyelids

*The major phakomatoses. From Geerates, W.J.: Ocular Syndromes. 3rd Ed. Philadelphia, Lea & Febiger, 1976, p. 613.

giomas of the choroid, retinal venous tortuosity, swelling of the optic nerve head, and optic atrophy have been observed. Phthisis and intraocular calcification may occur later in life.

ORBITAL-RETINAL-MIDBRAIN VASCULAR MALFORMATIONS

Arteriovenous malformations that involve the skin (scalp), orbit, retina, and midbrain have been described by Wyburn-Mason and Bonnet. This syndrome involves headache, seizures, mental retardation, and subarachnoid hemorrhage. The optic nerve may be occupied by a tangle of blood vessels.

RELATED TERMS

(1) Encephalotrigeminal angiomatosis; Sturge-Weber syndrome; nevus flammeus

(2) Midbrain-retinal angiomatosis; Bonnet's syndrome; Wyburn-Mason
(3) Neurofibromatosis; von Recklinghausen's disease
(4) Retinocerebral angiomatosis; von Hippel-Lindau disease
(5) Tuberous sclerosis; Bourneville's disease; epiloia

Vascular Anomalies

Some of the vascular anomalies that involve both the optic nerve and the brain have already been discussed in the section on phakomatoses. This section deals more specifically with isolated vascular anomalies that involve only the optic nerve. Persistence of elements of the hyaloid artery system provides another area of overlap with the previously described anomalies. Here we deal strictly with the posterior "vascular" component of the fetal vascular system.

Persistence of the Central Hyaloid Artery System. This condition is one of the most common developmental anomalies of the eye. The form and extent of persistence varies widely; part or all of the artery may persist from the disc to the lens. When the posterior end of the artery persists, attached to the disc, it may appear as a small bud of the central retinal artery as it emerges on the disc. More commonly, though, it takes the form of a single patent vessel or an obliterated strand arising from the artery and extending a variable distance into the vitreous. It may terminate as a fine strand, a bulb, or a heavy glial fan.

Prepapillary Vascular Loops. These are patent vessels (usually arterial) which form various loops and spirals extending from the disc (Fig. 3–7). They probably represent an irregular budding from the bulb of the hyaloid within Bergmeister's papilla, which remains after the rest of this structure atrophies. Prepapillary vascular loops most likely arise from a branch of the central retinal artery and are not connected directly with the fetal system of the tunica vasculosa. Venous loops are less common.

Abnormalities in the Branching of the Vessels. These abnormalities are common at the nerve level. Instead of two main divisions there may be multiple divisions, and usually the artery and the vein are similarly involved. Circular venous anastomoses may form on or around the disc with various types of communications. Various combinations of anomalous bifurcations and trifurcations may be seen.

Tortuosity of the Vessels. This condition may be localized to the nerve head or generalized throughout the retina, may occur unilaterally or bilaterally, and may involve arteries, veins, or both (Fig. 3–8). As a rule, the patient is asymptomatic. Corkscrew tortuosity of the vessels has been associated with congenital stenosis of the

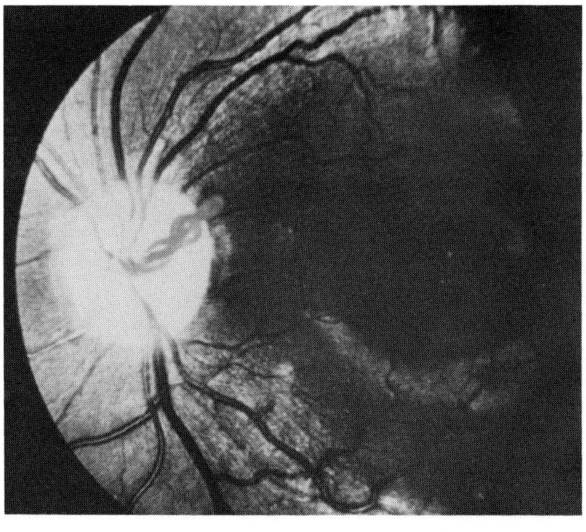

Fig. 3–7. An arterial prepapillary vascular loop arising as a branch of the central retinal artery.

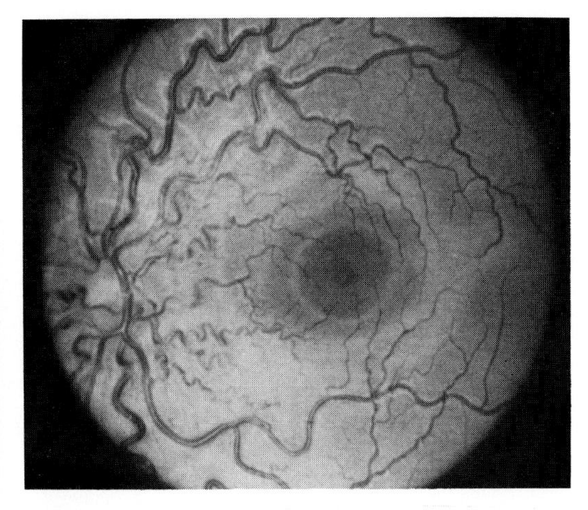

Fig. 3–8. Congenital tortuosity of retinal vessels. This condition can involve the arterial system, the venous system, or both systems and can occur unilaterally or bilaterally.

aorta. An unusual type of tortuosity in which the artery and vein twist around each other as they emerge from the disc has been described. This abnormal vascular pattern is also seen in association with other anomalies of the optic disc.

Abnormal Direction of the Emerging Retinal Vessels. This has been described as dysversion, or situs inversus, in Chapter 2, in the section on abnormal shape of the optic nerve.

Arteriovenous Anastomoses. This is an uncommon condition wherein the arteries connect directly with the veins without the interposition of a capillary network. It appears as a convoluted mass of vessels on the disc, so that the type or site of communication is impossible to assess (Fig. 3–9). This type of arteriovenous communication may arise in one of three ways: (1) There may be a direct end-on anastomosis between an artery or vein, (2) an angiomatous network of vessels may form in the anastomosis, or (3) an anomalous vessel may link the two. Similar involvement of the retinal vessels is mentioned in the discussion of von Hippel-Lindau disease.

Congenital Aneurysms. These aneurysms of the central retinal artery and of its main branches are rare and usually occur in otherwise healthy young people.

Abnormal Communications between the Retinal and Ciliary Systems. These anastomoses are relatively common and can be divided into three types: (1) Congenitally enlarged capillaries may run forward and anastomose on the surface of the disc, forming cilioretinal vessels. These vessels are derived from the circle of Haller-Zinn and usually extend from the temporal margin of the disc to the retina as an accessory artery. (2) Anastomoses from the peripapillary region of the choroid may run directly into the substance of the nerve to emerge at the margin of the disc and join the central retinal venous system as opticociliary vessels. (3) Anastomoses may run backwards from the choroid into the pial sheath of the nerve, forming choriovaginal veins.

RELATED TERMS

(1) Aneurysms
(2) Arteriovenous anastomosis
(3) Cilioretinal anastomoses
(4) Congenital tortuosity of the retinal vessels
(5) Hyaloid artery system persistence
(6) Prepapillary vascular loops

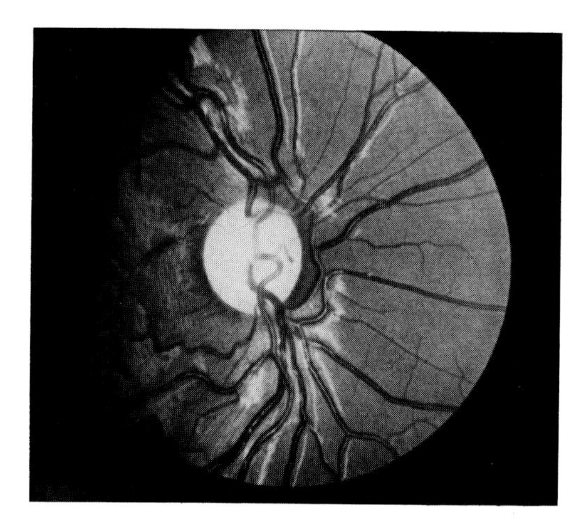

Fig. 3–9. Arteriovenous communication (anastomosis), showing a direct communication between the inferior temporal artery and the superior veins.

Albert, D.M.: General remarks on the phakomatoses. *In* Pathobiology of the Ocular Disease: A Dynamic Approach. Edited by A. Garner and G.K. Klintworth. New York, Marcel Dekker, 1982, Part A.

Alexander, G.L., and Norman, R.M.: The Sturge-Weber Syndrome. Bristol, John Wright & Sons, 1960.

Bailey, P.: Intracranial Tumors. Springfield, Charles C Thomas, 1933.

Bailey, P., and Cushing, A.: Classification of the Tumors of the Glioma Group. Philadelphia, W.B. Saunders, 1926.

Basse, H., and Schiffer, H.P.: Melanozytom der Papille. Ophthalmologica, *177*:245–247, 1978.

Byers, W.G.M.: Primary intradural tumors of the optic nerve. Studies Roy. Victoria Hosp., *1*:1–82, 1901.

Chutorian, A.M., Schwartz, J.F., Evans, R.A., and Carter, S.: Optic gliomas in children. Neurology, *14*:83–95, 1964.

Davis, W., and Thumin, M.: Cavernous hemangioma

of the optic disc and retina. Trans. Am. Acad. Ophthalmol. Otolaryngol., *60*:217–218, 1956.
Davis, F.A.: Primary tumors of the optic nerve. Arch. Ophthalmol., *23*:735–821, 957–1022, 1940.
Deutsch, A.R.: Pigmented tumor of the disc. Am. J. Ophthalmol., *58*:137–138, 1964.
deVeer, J.A.: Melanotic tumors of the optic nerve head. Arch. Ophthalmol., *65*:536–541, 1961.
Font, R., and Ferry, A.P.: The phakomatoses. Interntl. Ophthalmol. Clin., *12*:1–50, 1972.
Foos, R.Y., Straatsma, B.R., and Allen, R.A.: Astrocytoma of the optic nerve heads. Arch. Ophthalmol., *74*:319–326, 1965.
Ford, F.R.: Diseases of the nervous system. *In* Infancy, Childhood and Adolescence. 5th Ed. Springfield, Charles C Thomas, 1966.
Francois, J.: Ocular aspects of the phakomatoses. *In* Handbook of Clinical Neurology. Vol. 14. Edited by P.J. Vinken, and G.W. Bruyn. New York, Elsevier, 1972.
Francois, J., and Rabaey, M.: Tumeurs primitives du nerf optique. Acta Ophthalmol. (Copenh.), *30*:203–221, 1952.
Gass, J.D., and Braunstein, R.: Sessile and exophytic capillary angiomas of the juxtapapillary retina and optic nerve head. Arch. Ophthalmol., *98*:1790–1797, 1980.
Gass, J.D.M.: The Phakomatoses in Neuro-Ophthalmology. Edited by J.L. Smith. St. Louis, C.V. Mosby, 1965.
Goldsmith, J.: Neurofibromatosis associated with tumors of the optic papilla. Arch. Ophthalmol., *41*:718–729, 1949.
Griffith, A.D., and Sorsby, A.: The genetics of retinoblastoma. Br. J. Ophthalmol., *28*:279, 1944.
Hafner, G., and Meythaler, H.: A contribution to the clinical picture of phakomatoses. Ber. Dtsch. Ophthalmol. Ges., *74*:809–811, 1977.
Hall, G.S.: The ocular manifestations of tuberous sclerosis. Quart. J. Med., *15*:209–219, 1946.
Howard, G.M., and Ellsworth, R.M.: Differential diagnosis of retinoblastoma. Am. J. Ophthalmol., *60*:610–618, 1965.
Howard, G.M., and Forrest, A.W.: Incidence and location of melanocytomas. Arch. Ophthalmol., *77*:61–66, 1967.
Hoyt, W.F., and Beeston, D.: The Ocular Fundus in Neurolytic Disease. St. Louis, C.V. Mosby, 1966.
Isler, W.: Cerebrovascular diseases in the first three years of life. Brain Dev., *2*:95–105, 1980.
Juarez, C.P., and Tso, M.O.: An ultrastructural study of melanocytomas of the optic disk and uvea. Am. J. Ophthalmol., *90*:48–62, 1980.
Kramer, H.H., Karch, D., and Seibert, H.: Moyamoya-like vascular disease in tuberous sclerosis. Monatsschr. Kinderheilkd., *129*:595–597, 1981.
Lewis, R.A., and Riccardi, V.M.: Von Recklinghausen neurofibromatosis. Incidence of iris hamartoma. Ophthalmol., *88*:348–354, 1981.
Maki, Y., et al.: Computed tomography in Von Recklinghausen's disease. Childs. Brain, *8*:452–460, 1981.
Nicholson, D.H.: Tumors of the optic disc. Trans. Am. Acad. Ophthalmol. Otolaryngol., *83*:751–754, 1977.
Nielsen, P.G.: Capillary haemangioma of the optic disc. Acta Ophthalmol. (Copenh.), *57*:63–68, 1979.
Peterman, A.F., et al.: Encephalotrigeminal angiomatosis: A clinical study of 35 cases. J.A.M.A., *167*:2169–2176, 1958.
Rad, V.A., Sood, G.C., and Raman, R.: Ocular neurofibromatosis. Indian J. Ophthalmol., *29*:117–120, 1981.
Schwartz, P.L., Beards, J.A., and Morris, P.J.: Tuberous sclerosis associated with a retinal angioma. Am. J. Ophthalmol., *90*:485–488, 1980.
Silver, M.L.: Hereditary vascular tumors of the nervous system. J.A.M.A., *186*:1053–1056, 1954.
Slezak, H, Kenyeres, P., and Wiesflecker, J.: Differential diagnosis of the juxtapapillary racemose hemangioma. Klin. Monatsbl. Augenheilkd., *175*:306–308, 1979.
Stern, J., Jakobiec, F.A., and Housepian, E.M.: The architecture of optic nerve gliomas with and without neurofibromatosis. Arch. Ophthalmol., *98*:505–511, 1980.
Thomas, J.V., Schwartz, P.L., and Gragoudas, E.S: Von Hippel's disease in association with von Recklinghausen's neurofibromatosis. Br. J. Ophthalmol., *62*:604–608, 1978.
Thomas, M., and Burnside, R.M.: Von Hippel-Lindau disease. Am. J. Ophthalmol., *51*:140–146, 1961.
Tso, M.O., Zimmerman, L., and Fine, B.S.: The nature of retinoblastoma: Photoreceptor differentiation. Am. J. Ophthalmol., *69*:339, 1970.
Van der Hoeve, J.: Eye symptoms in phakomatoses. Trans. Ophthalmol. Soc. U.K., *52*:380–401, 1932.
Wallner, E.F., and Moorman, L.T.: Hemangioma of the optic disc. Arch. Ophthalmol., *53*:115–117, 1955.
Walsh, F.B.: The ocular signs of tumors involving the anterior visual pathways. Am. J. Ophthalmol., *42*:347–377, 1956.
White, J.P., and Loewenstein, A.: An unpigmented primary tumor of the optic disc. Br. J. Ophthalmol., *30*:253–260, 1946.
Wing, G.L., et al.: Von Hippel-Lindau disease: Angiomatosis of the retina and central nervous system. Ophthalmol., *88*:1311–1314, 1981.
Wyburn-Mason, R.: Arteriovenous aneurysm of midbrain and retina, facial naevi and mental changes. Brain, *66*:163–203, 1943.
Zaremba, J., et al.: Hereditary neurocutaneous angioma: A new genetic entity? J. Med. Genet., *16*:443–447, 1979.
Zaret, C.R., Choromokos, E.A., and Meisler, D.M.: Ciliooptic vein associated with phakomatosis. Ophthalmol., *87*:330–336, 1980.
Zimmerman, L.E.: Melanocytes, melanocytic nevi, and melanocytomas. Invest. Ophthalmol. Vis. Sci., *4*:11–41, 1965.

Section II

The Neuro-Ocular-Endocrine Dysplasia Syndromes

(Cephalad Neural Tube Dysgenesis)

Chapter 4

BASIC EMBRYOLOGY AND ANATOMY

EMBRYOLOGY

The cephalad neural tube develops early on (4-mm stage) into the prosencephalon (forebrain), mesencephalon (midbrain), and rhombencephalon (hindbrain) (Fig. 4–1).

At the 11-mm stage, the prosencephalon divides into the proximal diencephalon and the paired distal telencephalon; the mesencephalon develops into the supra-aqueductal tectum and the infra-aqueductal tegmentum.

The rhombencephalon further develops into the metencephalon (pons) and the myelencephalon (medulla) between the 11- and 16-mm stages. The cerebellum also arises at this time, from the metencephalic part of the rhombencephalon. Figure 4–2 illustrates fetal brain development between the 4- and 16-mm stages.

The telencephalon continues to expand into the cerebral hemispheres and lateral ventricles.

From the diencephalon arise the corpus callosum and septum pellucidum, hypophysis, thalamus, and hypothalamus.

The nuclei of the thalamus and hypothalamus develop from the walls of the diencephalon (13- to 18-mm stages) and progressively bulge into the cavity of the third ventricle, reducing it to a slit. The nuclei of the lateral geniculate body arise in this area between the 22- and 35-mm stages. At about the sixth month, they become laminated, which enables them to receive the fibers of the developing optic tract.

From the floor of the diencephalon, the hypophyseal diverticulum (infundibulum) arises, from which the stalk and the pars nervosa of the pituitary develop. Immediately cephalad is the thickening into which the optic nerves traverse to form the chiasm. The anterior lobe of the pituitary is determined by the ectodermal diverticulum of Rathke's pouch, which extends upward from the forward tip of the notochord towards the floor of the diencephalic part of the neural tube. The anterior and posterior lobes come into contact at the 14-mm stage and fuse to form the pituitary gland (Fig. 4–3).

Between the 16- and 22-mm stages, the anterior wall of the diencephalon forms the lamina reuniens, from which area the corpus callosum, septum pellucidum, and other midline structures later arise. Development of the septum pellucidum depends also on the normal expansion of the hemispheres and on development of the corpus callosum.

Any interruption of the developmental process during this induction stage may therefore lead to various combinations of morphologic and functional disturbances of the following:

(1) Cerebral hemispheres
(2) Lateral ventricles
(3) Superior colliculi

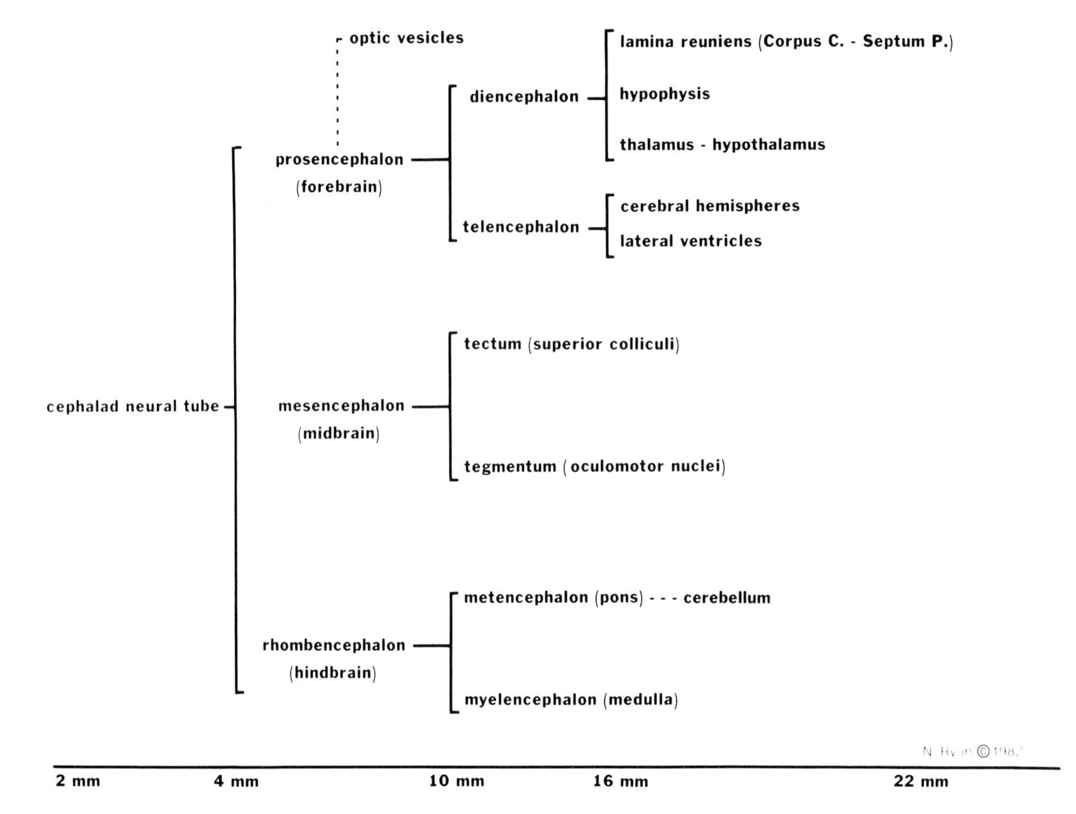

Fig. 4–1. Fetal brain. Developmental milestones of the cephalad neural tube.

(4) Oculomotor nuclei
(5) Corpus callosum; septum pellucidum
(6) Optic vesicles
(7) Hypophysis
(8) Thalamus; hypothalamus
(9) Cerebellum

ANATOMY

Functionally and anatomically the brain can be divided into the following structures:

(1) Cerebral cortex
(2) Basal ganglia
(3) Thalamus and hypothalamus
(4) Midbrain
(5) Brainstem: pons; medulla
(6) Cerebellum

The two cerebral hemispheres, which make up the largest portion of the brain, are joined across the longitudinal fissure by the great white central commissure, the corpus callosum. Each hemisphere may then by divided into the following:

(1) Frontal lobe: motor cortex
(2) Parietal lobe: somesthetic-stereognostic cortex
(3) occipital lobe: visual cortex
(4) Temporal lobe: auditory-olfactory-visual association cortex

The basal ganglia are masses of gray matter situated deep within the cerebral hemispheres. They function as part of the extrapyramidal system, which regulates motor integration. The corpus striatum, a component of the basal ganglia, consists of the following:

(1) Caudate nucleus
(2) Lentiform nucleus
    (a) Putamen
    (b) Globus pallidus
(3) Fascicles of internal capsule

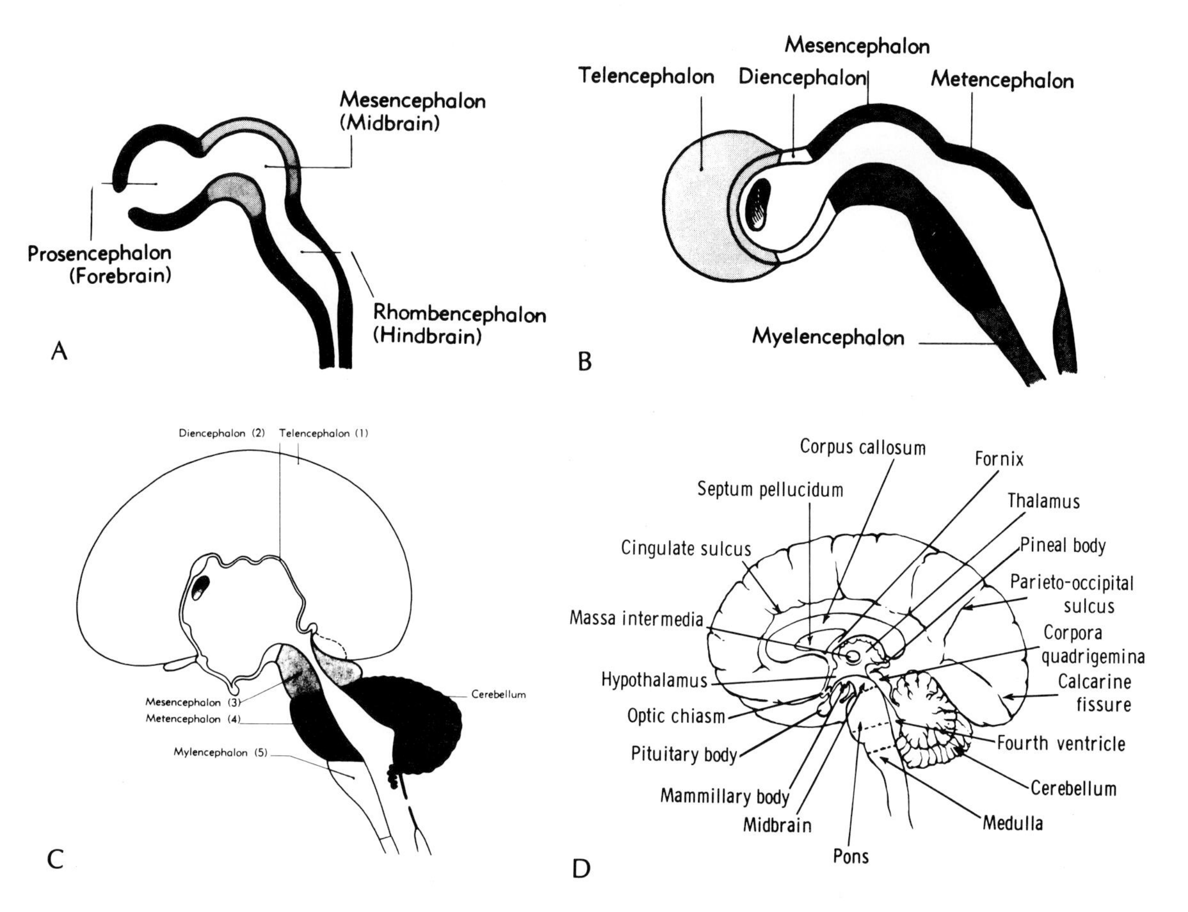

Fig. 4–2. Fetal brain development. *A.* 4 mm: shows the three major areas of the (1) forebrain (2) midbrain and (3) hindbrain. *B.* 11 mm: illustrates further division of the forebrain (telencephalon and diencephalon), midbrain (mesencephalon), and hindbrain (metencephalon and myelencephalon). *C.* 16 mm: illustrates further development of the cerebral cortex, midline structures, midbrain, pars, medulla and cerebellum. *D.* 12 wks: demonstrates completion of all major anatomical areas.

The thalamus is a large gray mass located on either side of the third ventricle and across the rostral end of the cerebral peduncles. On its lateral surface is the lateral geniculate body, which contains the synapse center for the retinal ganglion visual cells. The medial surface of the thalamus forms the lateral walls of the third ventricle and connects with the opposite thalamus by the massa intermedia. The thalamic radiations emerge from the lateral surface, enter the internal capsule, and terminate in the cerebral cortex for perception of somasthetic sensation.

The hypothalamus lies below (ventral to) the thalamus and forms the floor and inferolateral walls of the third ventricle. It includes the following:

(1) Mamillary bodies
(2) Tuber cinereum
(3) Infundibulum
(4) Optic chiasm

The infundibulum and the neural lobe of the hypophysis constitute the neurohypophysis.

The diverse functions of the hypothalamus include hormone release and stimulation, sexual drive, temperature control, emotional stability, and perhaps sleep.

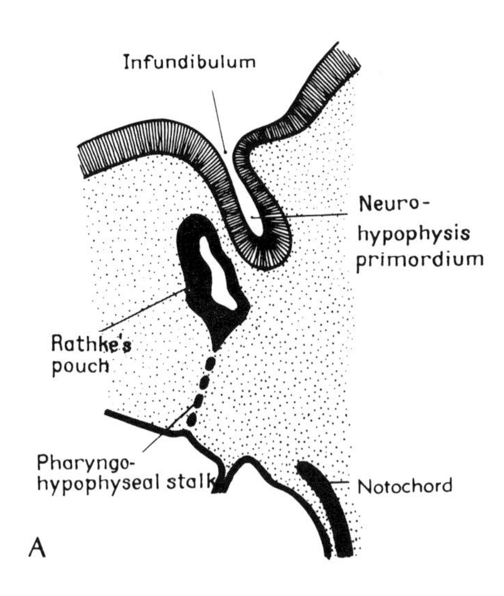

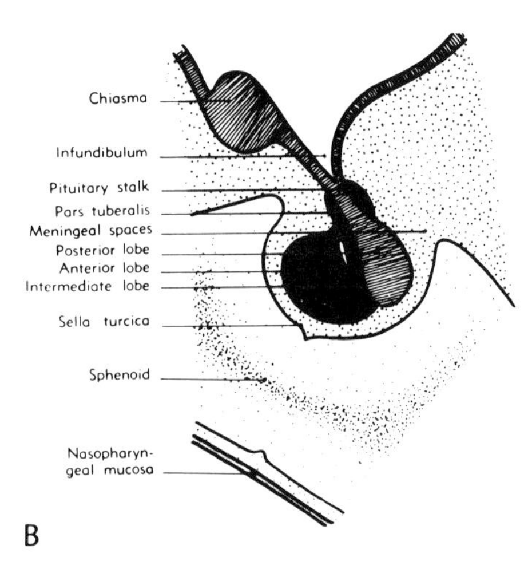

Fig. 4–3. Pituitary development. *A.* 6 mm: The infundibulum of hypophysis arises from the floor of the diencephalon and Rathke's pouch from the forward tip of the notochord.
B. 14 mm: The hypophyseal stalk, comprising the pars nervosa and Rathke's pouch, fuses with the anterior lobe to form the complete pituitary.

The short portion of the brain between the pons and the cerebral hemispheres is the midbrain, which consists of the dorsal (supra-aqueductal) tectum and the ventral (infra-aqueductal) tegmentum. The tectum contains the paired superior colliculi for visual reflexes and the paired inferior colliculi for auditory function; the tegmentum contains the oculomotor nuclei.

The brainstem is divided embryologically and anatomically into the pons and medulla, both of which develop from the rhombencephalon, or hindbrain.

The pons lies ventral to the cerebellum and anterior to the medulla. It is separated from the medulla by a groove through which the abducens, facial, and acoustic nerves emerge. The pons is demarcated anteriorly (cephalad) by the two cerebral peduncles, and posteroventrally by the middle cerebellar peduncle (brachium pontis).

The longitudinal fasciculi within the pons contain the following:

(1) Corticospinal tract
(2) Corticobulbar tract
(3) Frontopontine tract
(4) Parietotemporopontine tract

The cranial nerve nuclei in the pons include the abducens, facial, trigeminal, and auditory (cochlear and vestibular nerves).

The medulla oblongata is the portion of the brainstem between the spinal cord and the pons. The ventral part contains the central canal, and the dorsal part forms the floor of the fourth ventricle. The following subdivisions of the medulla may be recognized:

(1) Ventral section—contains the corticospinal tract
(2) Lateral section—contains the olive
(3) Dorsal section—contains the funiculus gracilis and cerebellum and funiculus cuneatus (internal arcuate fibers to the cerebellum and medial lemniscus)

Other important structures contained within the medulla include the posterior vestibular nuclear complex, hypoglossal nucleus, glossopharyngeal and vagal nuclei, olivary nuclei, spinocerebellar tract, and medial lemniscus.

Integration of reflexes concerned with swallowing, vomiting, respiration, and cardiovascular control occurs in the medulla oblongata.

The cerebellum, located in the posterior

fossa of the skull, is separated from the overlying cerebrum by dura (tentorium). It is composed of two large lateral masses (cerebellar hemispheres) and a median unpaired portion (the vermis).

The cerebellum contains three major projection bundles: the cerebellar peduncles. They consist of:

(1) Superior cerebellar peduncle (brachium conjunctivum)
 (a) Dentatorubral fibers
 (b) Ventral spinocerebellar tract
 (c) Uncinate fasciculus
(2) Middle cerebellar peduncle (brachium pontis): the pontocerebellar tract
(3) Inferior cerebellar peduncle (restiform body)
 (a) Olivocerebellar tract
 (b) Dorsal spinocerebellar tract
 (c) External arcuate fibers (medullocerebellar tract)

The cerebellum has several motor coordination functions, including (1) keeping the individual oriented in space (archicerebellum), (2) control of antigravity muscles (paleocerebellum), and (3) integration of volitional movements (neocerebellum).

Balinsky, B.: An Introduction to Embryology. 5th Ed. New York, Holt, Rinehart & Winston, 1981.
Brodal, A.: Neurological anatomy. In Relation to Clinical Medicine. 3rd Ed. New York, Oxford University Press, 1981.
Carpenter, M.B.: Human Neuroanatomy. 8th Ed. Baltimore, Williams & Wilkins, 1982.
Everett, N.B.: Functional Neuroanatomy. 6th Ed. Philadelphia, Lea & Febiger, 1971.
Gray, W.B., and Skandalkais, J.E.: Embryology for Surgeons. Philadelphia, W.B. Saunders, 1972.
Hamilton, W.J., and Mossman, H.W.: Human Embryology. 4th Ed. Baltimore, Williams & Wilkins, 1972.
Langman, J.: Medical Embryology. 4th Ed. Baltimore, Williams & Wilkins, 1981.
Martinez, and Martinez: Neuroanatomy, Philadelphia, W.B. Saunders, 1982.
Patten, B.M., and Carlson, B.M.: Foundations of Embryology. New York, McGraw-Hill, 1974.
Peters, A.: The Fine Structure of the Nervous System: The Neurons and Supporting Cells. Philadelphia, W.B. Saunders, 1976.
Truex, R.C., and Carpenter, M.B.: Human Neuroanatomy. 6th Ed. New York, State Mutual Bank, 1981.

Chapter 5

HOLOPROSENCEPHALY: THE OCULOCRANIOFACIAL SYNDROMES

Holoprosencephaly, which involves varying degrees of failure of interhemispheric cleavage, is the most frequently described entity of the congenital ocular-forebrain abnormalities. It is usually associated with craniofacial anomalies. Endocrinologic disturbances as well as dysgenesis of the deep cerebral venous system also occur. Arhinencephalia with anosmia, median cleft lip and palate, poikilothermy, hypotelorism, seizures, and psychomotor retardation have all been found in association with this form of cephalad neural tube dysgenesis.

The ocular abnormalities include microphthalmia, colobomata, retinal dysplasia, optic nerve hypoplasia, and even cyclopia in the more severe forms.

Several etiologic factors have been implicated in these ocular-forebrain malformations. They have been associated with maternal syphilis, diabetes mellitus, and toxoplasmosis. Abnormal chromosomal structure also has been noted in some reported cases. It is seen as a dominant form with incomplete penetrance, as a recessive form, or as a sporadic occurrence.

The pathogenesis of these disorders has been ascribed to failure of evagination of the secondary telencephalic vesicles with defective induction or development (fusion or cleavage) of the prosencephalon. Figures 5–1, 5–2, 5–3, and 5–4 demonstrate these prosencephalic defects.

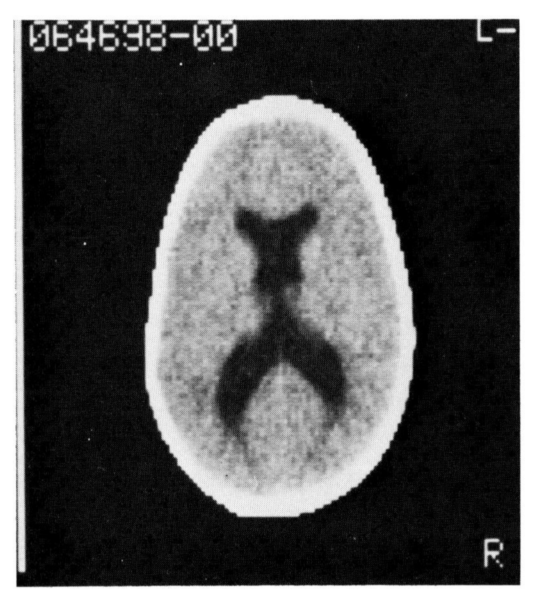

Fig. 5–1. Computed tomography demonstrating dysplasia of the ventriculoseptal system.

The prechordal mesoderm also is involved as it migrates forward anterior to the notochord during the third week, influencing the structural development of the forebrain and midline components of the face.

The variety of clinical expressions of these developmental defects has given rise to a

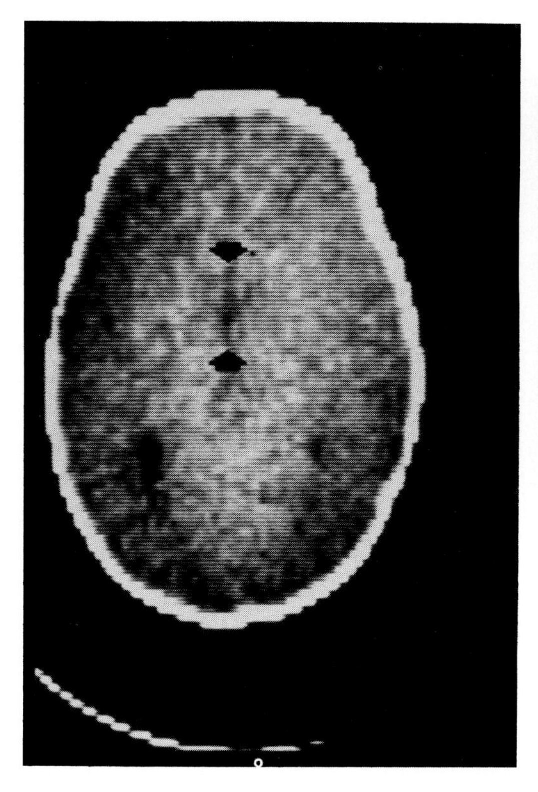

Fig. 5–2. Computed tomography showing aplasia of corpus callosum.

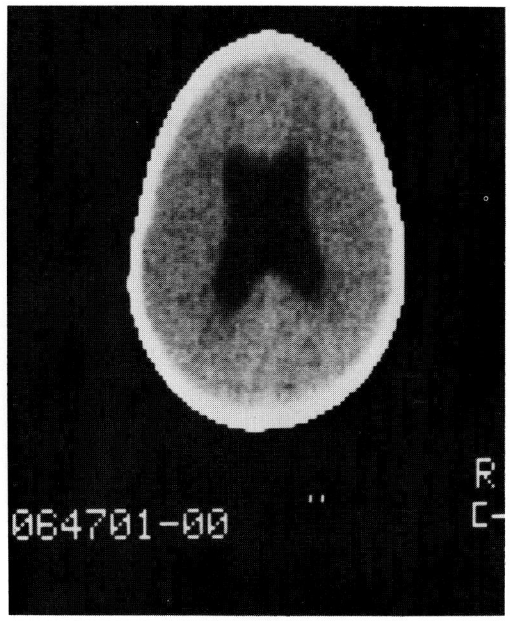

Fig. 5–3. Computed tomography showing absence of septum pellucidum.

wide spectrum of related "neuro-ocular-endocrine" syndromes.

Another clinical expression of the holoprosencephalic spectrum is the association of craniofacial disorders with the tilted optic disc (Fig. 5–5) as well as with situs inversus of the disc, heterotopia of the maculae, and hypopigmentation of the ocular fundus. Associated anomalies include:

(1) Orbital hypertelorism
(2) Crouzon's disease with oxycephaly
(3) Apert's disease with syndactyly
(4) Midfacial hypoplasia: frontonasal dysplasia

Another related group of oculocraniofacial defects involves the association of ocular anomalies with basal encephalocele, including hypertelorism, exotropia, microphthalmus, cryptophthalmus, peri-

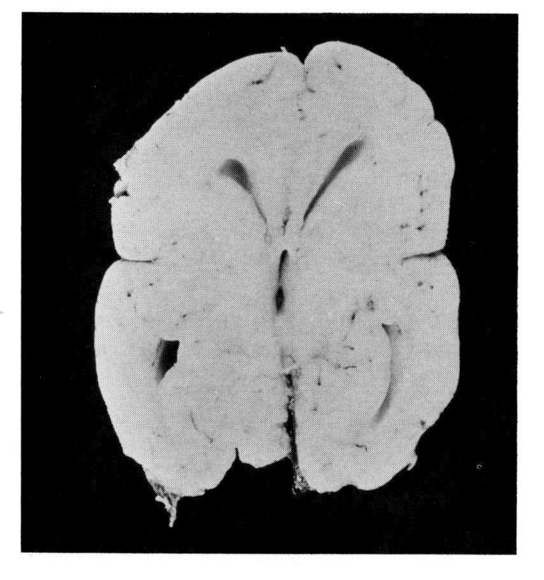

Fig. 5–4. Gross brain section showing ventriculoseptal dysplasia.

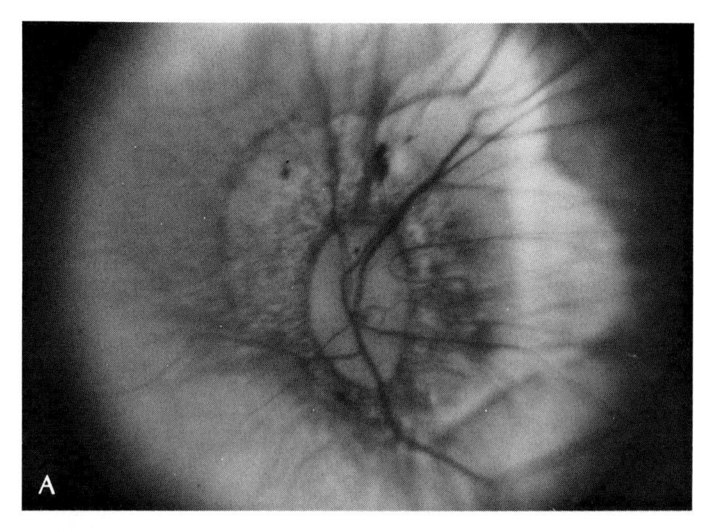

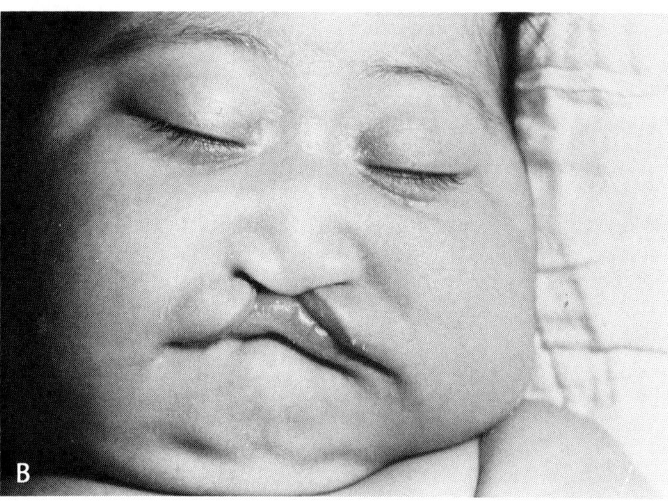

Fig. 5–5. *A,* Optic nerve dysplasia showing a tilted nerve with dysversion of the nerve and a posterior staphyloma. *B,* Midline cleft deformity in the same patient.

papillary staphyloma, optic nerve pallor, pits, colobomas, and megalopapilla.

Cockayne's syndrome displays another rare combination of neuro-ocular-endocrine and skeletal defects which include dwarfism, prominent maxilla, osteoporosis, microcephaly, deafness, joint contractures, hyperreflexia, and early death from inanition and infection. Retinal vascular anomalies, optic nerve pallor, retinal degeneration, and intracerebral calcification have been noted.

The Kenny syndrome consists of dwarfism, tetany with transient hypocalcemia, and retarded bone age, along with prominent forehead and micrognathia. The ocular findings consist of nanophthalmus, microphthalmus, pseudopapilledema, drusen, congenital tortuosity of the retinal vessels, and macular changes. This syndrome may represent a variant of congenital hypoparathyroidism with abnormal calcium metabolism.

Apple, D.J.: Chromosome induced ocular disease. *In* Genetic and Metabolic Eye Disease. Edited by M.F. Goldberg, Boston, Little, Brown, 1974.

Awan, K.J.: Hypotelorism and optic disc anomalies. Ann. Ophthalmol., *9*:771–777, 1977.

Buhler, E., et al.: Trisomy 13–15 mit cebocephalie. Annales de Pediatre, *199*:198–205, 1962.

Cockayne, E.A.: Dwarfism with retinal atrophy and deafness. Arch. Dis. Child., *11*:11–18, 1936.

Cohen, M.M., Jr., et al.: Holoprosencephaly: Nosology, etiology and pathogenesis. Birth Defects, *7*:125–135, 1971.

Dallaire, L., Fraser, F.C., and Wiglesworth, F.W.: Familial holoprosencephaly. Birth Defects, *7*:136–142, 1971.

DeMyer, W.: A 46-chromosome cebocephaly, with remarks on the relation of 13–15 trisomy to holoprosencephaly (archinencephaly). Annales de Pediatre, *203*:169–177, 1964.

Francois, J.: Genetic aspects of ophthalmology-autosomal chromosome aberrations in ophthalmology. Int. Ophthalmol. Clin., *8*:839–910, 1968.

Goldhammer, Y., and Smith, J.L.: Optic nerve anomalies in basal encephalocele. Arch. Ophthalmol., *93*:115–118, 1975.

Greenfield, P.S., et al.: Hypoplasia of the optic nerve in association with prosencephaly. J. Pediatr. Ophthalmol. Strabismus, *15*:222–225, 1978.

Hintz, R.L., Menking, M., and Sotos, J.F.: Familial holoprosencephaly with endocrine dysgenesis. J. Pediatr., *72*:81–87, 1968.

Kenny, F.M., and Linarelli, L.: Dwarfism and cortical thickening of tubular bones. Am. J. Dis. Child., *111*:201–207, 1966.

Margolis, S., and Siegel, I.M.: The tilted disc syndrome, *in* cranio-facial diseases. *In* Neuro-Ophthalmology Focus. Edited by J.L. Smith. New York, Masson, 1980.

Smith, D.W.: Recognizable Patterns of Human Malformation: Genetic, Embryologic and Clinical Aspects. Vol. 7. 2nd Ed. Philadelphia, W.B. Saunders, 1976.

Chapter 6

OPTIC NERVE HYPOPLASIA: SEPTO-OPTIC-PITUITARY DYSPLASIA*

For almost a century (1864 to 1962), optic nerve hypoplasia was described as a rare and isolated ocular anomaly characterized by a small, pale optic nerve and blindness.

Only during the last 20 years, and essentially during the last decade, has optic nerve hypoplasia been described in association with congenital defects of certain anterior midline brain structures—notably, the septum pellucidum and pituitary. Also, only during this later period has the wide disparity in the clinical expressions of these related defects been emphasized and the entire spectrum of the "septo-optic-pituitary dysplasia" syndrome been fully recognized.

With the advent of ultrasonography and computed tomography, size of the optic nerves can be determined, and the presence of associated midline brain defects can be detected in a noninvasive and atraumatic manner.

The purpose of this report is to study a large group of patients with optic nerve hypoplasia quantitatively to determine more precisely the degree of dysplasia (hypoplasia), its relationship to visual function, and the frequency of coincidence with anterior midline brain defects and dysfunction. A further purpose is to review possible teratogenic, intrauterine, environmental, and genetic factors that may have a pathogenic role in the development of these correlated congenital defects. Development of a cohesive classification system for the septo-optic-pituitary dysplasia syndrome is a final purpose of this report.

HISTORICAL BACKGROUND

In 1884, Magnus described a small, pale optic nerve, with presence of the retinal vessels, in one eye of a small child. The child had no apparent useful vision in that eye. This case presentation was the first to describe completely all characteristics of optic nerve hypoplasia. During the next 50 years, 15 isolated case reports of optic nerve hypoplasia were added to the medical literature.

In 1941, Scheie and Adler described complete aplasia (absence) and partial aplasia (hypoplasia) of the optic nerves. They decribed a 3-year-old male with congenital blindness, fixed and unreactive pupils, wandering nystagmus, and small, pale optic discs with normal retinal vasculature. They proposed that the embryogenesis was a fetal developmental arrest of the primary mesoderm in aplasia and of the ganglion cell layer in hypoplasia.

Reeves was the first to report the asso-

*Abstracted from the author's American Ophthalmological Society Thesis, Trans. Am. Ophthalmol. Soc., *79*:425–457, 1981.

ciation of optic nerve hypoplasia and absence of the septum pellucidum. In 1941, he described these conditions in a 7-month-old infant with congenital blindness but otherwise normal development. The pupillary reflexes were absent, and the optic nerves were small, pale, and "aplastic." Absence of the septum pellucidum was demonstrated by pneumoencephalography.

In 1956, De Morsier described the necropsy findings of a patient with hypoplasia of the optic nerves and agenesis of the septum pellucidum, which he defined as "septo-optic dysplasia." Within six years, he accumulated 36 cases of septum pellucidum agenesis, nine of which demonstrated an associated optic nerve hypoplasia.

Two reports that appeared in the literature in 1970 described 45 new cases of optic nerve hypoplasia, which indicate that the condition was not as rare as it was previously believed to be.

The association of pituitary dwarfism with septo-optic dysplasia was first described by Hoyt, Kaplan, Grumback, and Glaser, who in 1970, reported nine patients with these associated findings. Pneumoencephalography was performed on three of the children, and absence of the septum pellucidum was demonstrated in all three.

In 1978, Clark and Meyer emphasized the importance of recognizing neonatal hypoglycemia and seizures in association with congenital blindness as characteristics of septo-optic-pituitary dysplasia in infancy.

Segmental Optic Nerve Hypoplasia

Schwartz's report in 1915 prompted allusions to the concept of segmental (partial or incomplete) optic nerve hypoplasia. He described a patient with bilateral optic nerve hypoplasia, 20/40 vision, and binasal hemianopic field defects. He postulated an isolated failure in development of the uncrossed nerve fibers.

In 1972, Seely and Smith reviewed 12 cases of optic nerve hypoplasia with documented visual field defects. Seven of these cases were bilateral, with binasal field defects in four, nasal field defects in one, temporal field defects in one, and generalized constriction in one. Field defects demonstrated by the five unilateral cases were nasal in three, temporal in one, and central in one. In four cases of their own, three were bilateral with centrocecal defects; the unilateral case showed temporal field loss.

Tilted Disc Syndrome

Rucker, in 1946, described six patients exhibiting the tilted disc-conus-ectasia of the "choroid" with bitemporal field defects (Fig. 6–1). He emphasized that the field defects often extended somewhat into the upper nasal quadrants and thus were distinguishable from true bitemporal defects, which result from chiasmal lesions.

Petersen and Walton, in 1977, speculated on the relationship of the "tilted disc syndrome" to segmental optic nerve hypoplasia. They described 17 children, all born of diabetic mothers, who exhibited segmental optic nerve hypoplasia with normal visual acuity and visual field defects corresponding to the hypoplastic area of the disc. Figure 6–2 shows the various types of optic nerve hypoplasia.

In 1978, Dorrell studied 60 tilted discs,

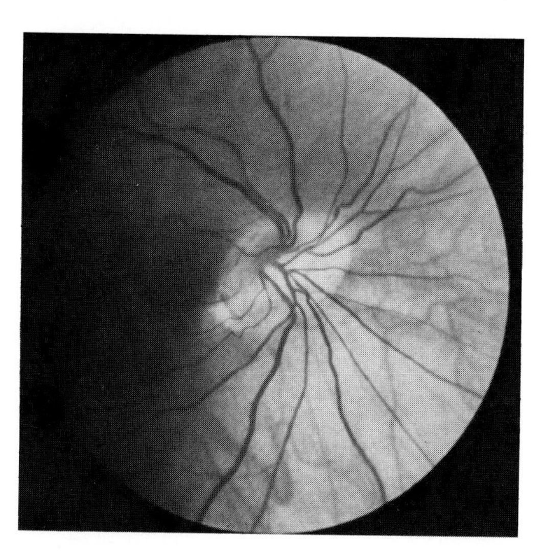

Fig. 6–1. Tilted disc-conus-ectasia, with temporal visual field defects.

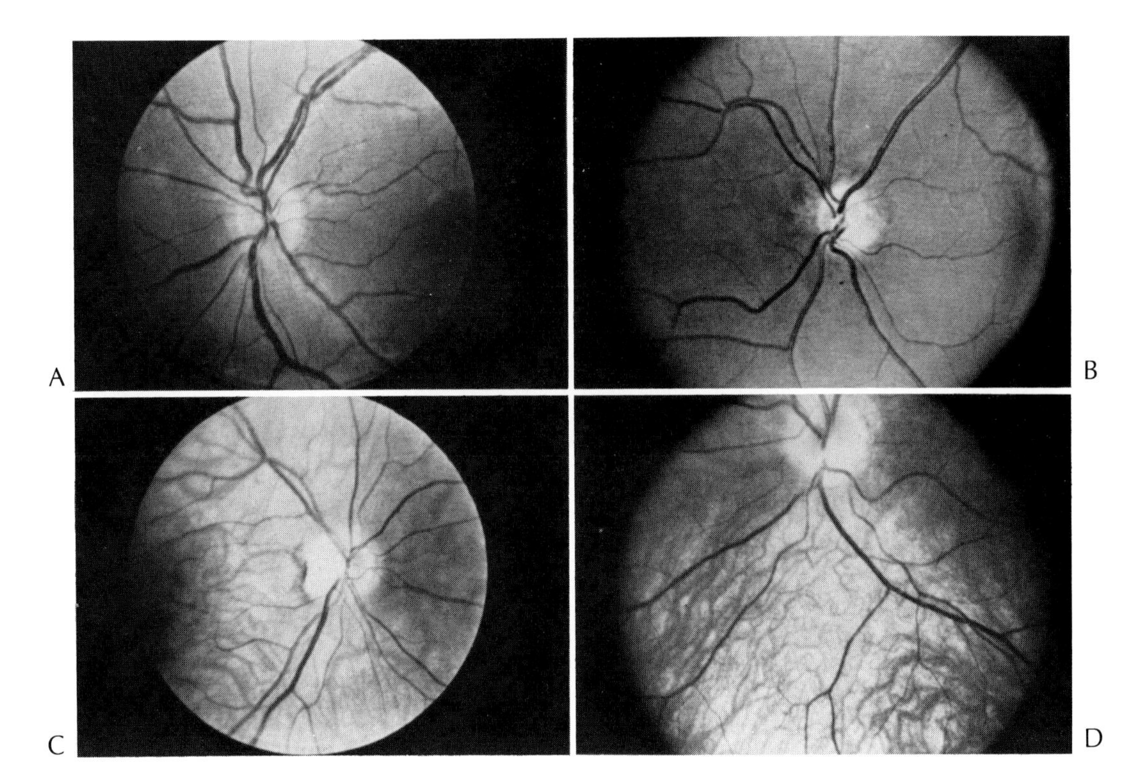

Fig. 6–2. Optic nerve hypoplasia. *A,* Normal. *B,* Diffuse hypoplasia. *C,* Segmental hypoplasia. *D,* Tilted disc-conus-ectasia.

each demonstrating the crescent, the depression on one side, and the oblique direction of the retinal vessels as they entered the eye. The direction of the tilt was nasal in 36 eyes (60%), downward in 7 (12%), and inferotemporal in 17 (28%). He found that the visual field defects corresponded to the direction of the tilt and suggested that this correspondence provided further evidence of diminished axons.

EMBRYOGENESIS

The embryogenesis of optic nerve hypoplasia is uncertain and perhaps as multifarious as its clinical expression. Within the framework of currently accepted developmental stages, three major embryogenic theories have evolved.

According to one of the theories, when the optic nerve hypoplasia occurs as an isolated ocular defect, primary failure of the ganglion cells occurs between the 12- and 17-mm stages (Fig. 6–3).

The other two embryonic theories concern the occurrence of optic nerve hypoplasia in association with central nervous system anomalies. The first theory suggests that normally developing ganglion cells reach a malformed chiasm at the 18-mm stage, cannot proceed across the midline into the opposite tracts, and undergo retrograde degeneration thereafter. The second theory suggests that the hypoplasia results from stretching of the optic nerves during abnormal development of the cerebral hemispheres and ventricular system, a process that results in retrograde degeneration of the ganglion cells of the retina. This theory further proposes that when mainly uncrossed fibers are affected, developmental arrest may occur even later, since normally, these fibers grow along the

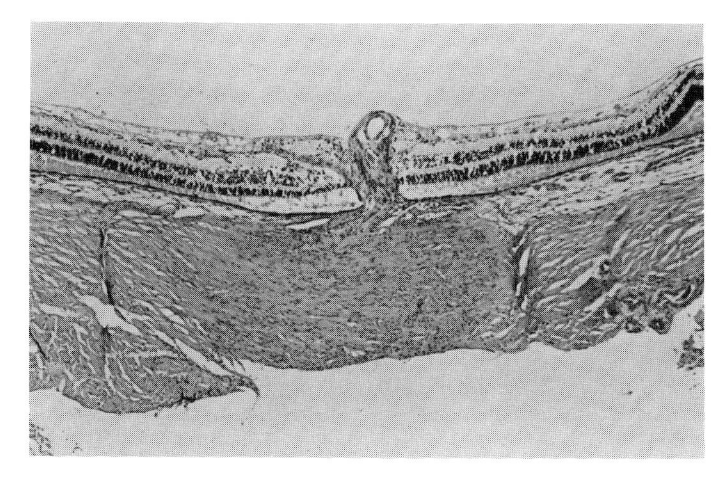

Fig. 6–3. Retinal ganglion cell dysplasia seen in microscopic section of retina in optic nerve hypoplasia.

optic nerve at a later time than the crossed fibers do.

HISTOPATHOLOGY

In 1979, Hotchkiss and Green presented the clinical and histopathologic features of 22 cases of optic nerve hypoplasia (involving 35 eyes). Autopsy data was correlated in 18 of the 22 cases, and in the remaining four instances, the eyes were surgically obtained. Seventeen of the 18 autopsy cases had involved stillbirths, 12 of which were anencephalic.

The configuration of the optic nerve head varied considerably. In some, the retinal pigment epithelium extended over the edge of the nerve head with a prominent scleral-laminar junction and formed the "double ring sign." In others, there was normal termination of the retinal pigment epithelium at the disc margin.

The nerve fiber content of the optic nerves also varied. In 77% of the eyes studied, there were only glial elements, in varying numbers. The other 23% showed the presence, but reduced number, of nerve fibers in the optic nerve. Retinal ganglion cells were absent in 54% of the eyes and present in reduced number in 40%. One bilateral case showed a nearly normal complement of ganglion cells. In 66% of the eyes, no nerve fiber layer could be detected, and nerve fibers in the remaining eyes were present in reduced number. The retinal vessels were considered to be normal in 80% of the eyes and less than a full complement in 20%. A posterior choroidal coloboma was present in 17% of the eyes. Persistent remnants of the hyaloid vascular system were present in 37% of the eyes.

The maternal history revealed an incidence of hydramnios in 33%, significant illness during pregnancy in 33%, and a history of drug exposure in 22%. The median maternal age was 25 years. Thirty-three percent of the mothers were primiparae. Sixty-four percent of the 22 cases involved females, and 64% of the cases were bilateral.

PATHOGENESIS

Little is known about specific causative factors responsible for the developmental defect in optic nerve hypoplasia. Two primary modes of transmission have been proposed as possible causative factors: (1) a genetic alteration, and (2) alteration of the intrauterine environment by maternal stress, either metabolic or toxic.

A. Genetic Factors:
   1. Familial occurrences
   2. Young maternal age
   3. Firstborn child

B. Alteration of Intrauterine Environment:
   1. Maternal drug ingestion
   2. Maternal diabetes mellitus
   3. Maternal-fetal cytomegalovirus infections

THE ACERS STUDY

The 45 patients recalled for the study were obtained from a total of 63 patients with a recorded diagnosis of optic nerve hypoplasia, from within a population base of 3,000,000—an approximate overall incidence of 2:100,000 population.

In this patient group, the ratio of bilateral to unilateral occurrence was 40:5, and the ratio of male to female occurrence was 23:22. Age ranged from one month to 59 years.

Maternal History (40 Patients).* The mean maternal age was 23.8 years, compared with a mean maternal age of 24.6 years in the general population of the state. Twenty-one of the 40 patients (52%) were firstborn children.

Five of the 40 patients were born to diabetic mothers, an incidence of 12.5%. Significant illnesses or specific complications of pregnancy occurred in six cases (15%), including pre-eclampsia (two cases), hypertension (two cases), and kidney infection (two cases).

An admitted history of drug ingestion was reported in five cases (12.5%). In only one case though was the ingestion of an abortive agent recorded; the origin in that instance was possibly quinine.

Visual Function. Twenty-one of the 45 patients were infants or preliterate children in whom only gross estimation of visual function could be made. Ten of these 21 patients (47%) appeared to have bilateral involvement with no light perception (response) or observable visual attention. Eight patients with bilateral involvement demonstrated varying degrees of visual attention or light responsiveness. Two patients demonstrated unilateral optic nerve hypoplasia with evidence of marked amblyopia in the involved eye. One patient had bilateral optic nerve hypoplasia with visual attention (response) in one eye only.

There were 24 patients in the literate age group. Visual acuity ranged from 20/30 to no light perception. Three of the 24 patients had no light perception in either eye. Three patients had unilateral optic nerve hypoplasia with 20/20 vision in the normal eye and no light perception in the involved eye. Visual field studies were obtained on 17 of these patients and demonstrated bitemporal defects in seven, generalized constriction in six, central defects in one, binasal defects in one, and altitudinal defects in two. In the three unilateral cases, the visual fields in the uninvolved eye were normally full, and in two of these patients, the visual fields in the involved amblyopic eyes were unobtainable. The third patient showed a central defect in the involved eye.

Optic Nerve Measurements. Echographic measurements of the horizontal and vertical diameters of the optic nerves were obtained, and the area of the optic nerve was calculated for all 45 patients. The echographic areas ranged from 2.84 sq mm (decimal of normal 0.39) to 6.74 sq mm (decimal of normal 0.93) with a mean echographic area of 4.61 sq mm (decimal of normal 0.63). The nine patients with tilted nerves usually demonstrated an optic nerve area greater than that of the general group of optic nerve hypoplasia patients, but still less than a normal area. This group ranged from 4.41 sq mm area (decimal of normal 0.61) to 6.74 sq mm (decimal of normal 0.75). The five unilateral cases had a mean echographic area of 5.12 sq mm (decimal of normal 0.71) in the involved eye and an echographic area of 7.57 sq mm (decimal of normal 1.0) in the normal eye.

Computed Axial Tomography of the Midline Brain. Computed axial tomographic studies of the midline brain structures were performed on all 45 patients. Twelve patients (27%) demonstrated midline anomalies, specifically, partial or com-

*Maternal histories for five patients were unobtainable.

plete absence of the septum pellucidum. Three of the 12 also had other midline defects, including diverticulum of the third ventricle, optic vesicle cyst, and diffuse dysgenesis. None of the nine patients with tilted optic nerve syndrome demonstrated a midline brain anomaly.

Associated Neuroendocrine Dysfunction. Six of the 45 (13%) patients demonstrated associated neurological defects, including epilepsy (two patients), cerebral palsy (two patients), and mental retardation (two patients).

Six other patients in the study group demonstrated clinical evidence of pituitary hypofunction, including growth retardation in three patients and neonatal hypoglycemia with seizures in three patients. Four of the six patients received further neuroendocrine testing, and all demonstrated pituitary hypofunction, with diminished growth hormone (two patients), multiple trophic hormones (four patients), and hypoglycemia (two patients). One other patient, a 15-year-old, demonstrated borderline pituitary function but definite growth retardation. One of the six patients died before neuroendocrine studies were completed. Computed axial tomography had demonstrated partial or total absence of the septum pellucidum in all six patients.

Associated Ocular Anomalies. These included nystagmus, strabismus, colobomas, latent nystagmus, Duane's syndrome, and aniridia.

CLASSIFICATION OF THE OPTIC NERVE HYPOPLASIA SYNDROME

Data from this study indicate that 12 out of 45 patients with optic nerve hypoplasia showed computed tomographic evidence of partial or complete absence of the septum pellucidum—an incidence of 27%.

Echographic measurements of the tilted nerve syndrome patients indicate that they, too, demonstrated structural hypoplasia of the optic nerves and visual dysfunction. In this group of ninc paticnts, cchographic areas of the optic nerve ranged from 61% to 93% of the normal area. The echographic diameter of the nerve measured less in the direction of the tilt. The visual acuities in this group were generally good. The visual field defects almost always corresponded to the direction of the tilt.

These data suggest that optic nerve hypoplasia not only is a component of a more diffuse syndrome, but can itself be subdivided into clinical subtypes.

Optic nerve hypoplasia is not a rare and isolated ocular anomaly, therefore, but part of a spectrum of related development anomalies that affect the optic nerves and other structures in the cental nervous system. It is a syndrome with disparate clinical expressions.

The correlated embryologic and clinical factors that involve the optic nerves, septum pellucidum, and hypophyseal-pituitary axis provide common ground for a classification system, which is supported by the data in this report.

THE OPTIC NERVE HYPOPLASIA SYNDROME

Type I.	Optic Nerve Hypoplasia Simplex
	A. Diffuse
	B. Segmental
Type II.	Septo-Optic Dysplasia
Type III.	Septo-Optic-Pituitary Dysplasia

Acers, T.E.: Optic nerve hypoplasia: The septo-optic-pituitary syndrome. Trans. Am. Ophthalmol. Soc., *79*:425–457, 1981.

Clark, E.A., and Meyer, W.F.: Blindness and hypoglycemia: Growth hormone deficiency with septo-optic dysplasia. Tex. Med., *74*:47–50, 1978.

DeMorsier, G.: Median cranioencephalic dystraphias and olfactogenital dysplasia. World Neurology, *3*:485, 1962.

Dorrell, D.: The tilted disc. Br. J. Ophthalmol., *62*:16–20, 1978.

Edwards, W.C. and Layden, W.E.: Optic nerve hypoplasia. Am. J. Ophthalmol., *7*:950–959, 1970.

Elster, A.B., and McAnarney, E.R.: Maternal age regarding septo-optic dysplasia. J. Pediatr., *94*:162, 1972.

Fuchs, E.: Uber den anatomischen befund einiger angeborenen anomalien der netzhaut und des sehnerven. Albrecht Von Graefes Arch. Klin. Exp. Ophthalmol., *28*:139, 1882.

Hackenbruch, Y., et al.: Familial bilateral optic nerve hypoplasia. Am. J. Ophthalmol., *79*:314–320, 1975.

Hittner, H.M., Desmond, M.M., and Montgomery, J.R.: Optic nerve manifestations of cytomegalovirus infection. Am. J. Ophthalmol., *81*:661–665, 1976.

Hotchkiss, M., and Green, W.R.: Optic nerve aplasia and hypoplasia. J. Pediatr. Ophthalmol., *16*:225–240, 1979.

Hoyt, C.A., and Billson, F.A.: Maternal anticonvulsants and optic nerve hypoplasia. Br. J. Ophthalmol., *62*:3–6, 1978.

Hoyt, W.F., et al.: Septo-optic dysplasia and pituitary dwarfism. Lancet, *1*:893–894, 1970.

Kraus-Brucker, W., and Gardner, D.W.: Optic nerve hypoplasia associated with absent septum pellucidum and hypopituitarism. Am. J. Ophthalmol., *89*:113–120, 1980.

Lippe, B., Kaplan, S.A., and La Franchi, S.: Septo-optic dysplasia and maternal age. Lancet, *2*:92, 1979.

Magnus, H.: Zur karuistik der angeborenen sehnerven-missildungen. Klin. Monatsbl. Augenheilkd., *2*:85–87, 1884.

Margolis, S., and Siegel, I.M.: The tilted disc syndrome in craniofacial diseases. *In* Neuro-Ophthalmology Focus. Edited by J.L. Smith. New York, Masson, 1980.

McKinna, A.J.: Quinine induced hypoplasia of the optic nerve. Can. J. Ophthalmol., *1*:261–263, 1966.

Missiroli, G.: Una nuova syndrome congenita a carattere famigliore: Ipoplasia del nervo optico ed emianopsia binasale. Boll. Ocul., *26*:683–698, 1947.

Petersen, R.A., and Walton, D.S.: Optic nerve hypoplasia with good visual acuity and visual field defects. Arch. Ophthalmol., *95*:254–258, 1977.

Reeves, D.L.: Congenital absence of the septum pellucidum. Bull. Johns Hopkins Hosp., *69*:61–71, 1941.

Rucker, C.W.: Bitemporal defects in the visual fields resulting from developmental anomalies of the optic discs. Arch. Ophthalmol., *35*:546–554, 1946.

Scheie, H.C., and Adler, F.H.: Aplasia of the optic nerve. Arch. Ophthalmol., *26*:61–70, 1941.

Shwarz, O.: Ein Fall von manglhafter Bilding (Hypoplasie) beider sehnerven. Arch. f. Ophth., *90*:326–328, 1915.

Seely, R.L., and Smith, J.L.: Visual field defects in optic nerve hypoplasia. Am. J. Ophthalmol., *73*:882–889, 1972.

Van Dyk, H.J.L., and Morgan, K.S.: Optic nerve hypoplasia and young maternal age. Am. J. Ophthalmol., *89*:879, 1980.

Von Szily, A.: Uver den conus in heterotypisher richtung. Albrecht Von Graefes Arch. Klin. Exp. Ophthalmol., *110*:183–291, 1922.

Walton, D.S., and Rabb, R.M.: Optic nerve hypoplasia. Arch. Ophthalmol., *84*:572–578, 1970.

Ziering, J.: Der papillare und parapapillare conus heterotypicus. Klin. Monatsbl. Augenheilkd., *97*:169–184, 1936.

Chapter 7

CONGENITAL OPHTHALMOPLEGIA PLUS

Recent advances in computed tomography have given rise to newly revealed syndromes that are clinically and developmentally related. Several reports on correlated studies of cephalad neural tube defects have further advanced the spectrum of neuro-ocular-endocrine syndromes.

One recent study describes concomitant forebrain-midbrain developmental anomalies in three patients, including:

(1) Bilateral congenital ophthalmoplegia
(2) Corpus callosum/septum pellucidum dysplasia
(3) Optic nerve dysplasia
(4) Hypopituitarism

Patients with these anomalies have congenital bilateral ophthalmoplegia in common and thus have been described as having "congenital ophthalmoplegia plus" syndrome. Two patients in the study had levator involvement with ptosis, and all three had normal pupils. Two patients had optic nerve dysplasia, and two patients had dysplasia of the corpus callosum and lateral ventricles. One patient showed evidence of hypopituitarism with a growth hormone deficiency.

Congenital ophthalmoplegia usually has been reported in heredofamilial and incomplete forms and has been associated with nerve deafness, long tract involvement, and limb-girdle myopathy.

Defects associated with congenital corpus callosum dysgenesis have also been reported, including epilepsy, optic nerve hypoplasia, craniofacial dystraphicus, mental retardation, neuroendocrine defects, and long tract dysfunctions.

The acquired, or developmental, ophthalmoplegia plus syndrome is distinct and includes progressive external ophthalmoplegia with spinocerebellar ataxia, limb-girdle myopathy, heart block, cardiomyopathy, mitral valve prolapse, neural deafness, retinitis pigmentosa, and aminoacidopathy syndromes (Fig. 7–1).

Isolated congenital bilateral horizontal gaze palsy also has been described; the proposed etiologies have ranged from a bilateral supranuclear defect due to aplasia of the paramedian zone of the pontine tegmentum to primary myopathy of the extraocular muscles.

Several cases describing bilateral external ophthalmoplegia with retention of levator activity have been reported (Fig. 7–2). Bilateral ophthalmoplegia with levator involvement and pupil sparing has also been reported. Total ophthalmoplegia with ocular immobility, ptosis, and pupillary involvement is the rarest form of congenital ophthalmoplegia. The association of congenital ophthalmoplegia with dysgenesis of the corpus callosum, dysplasia of the optic

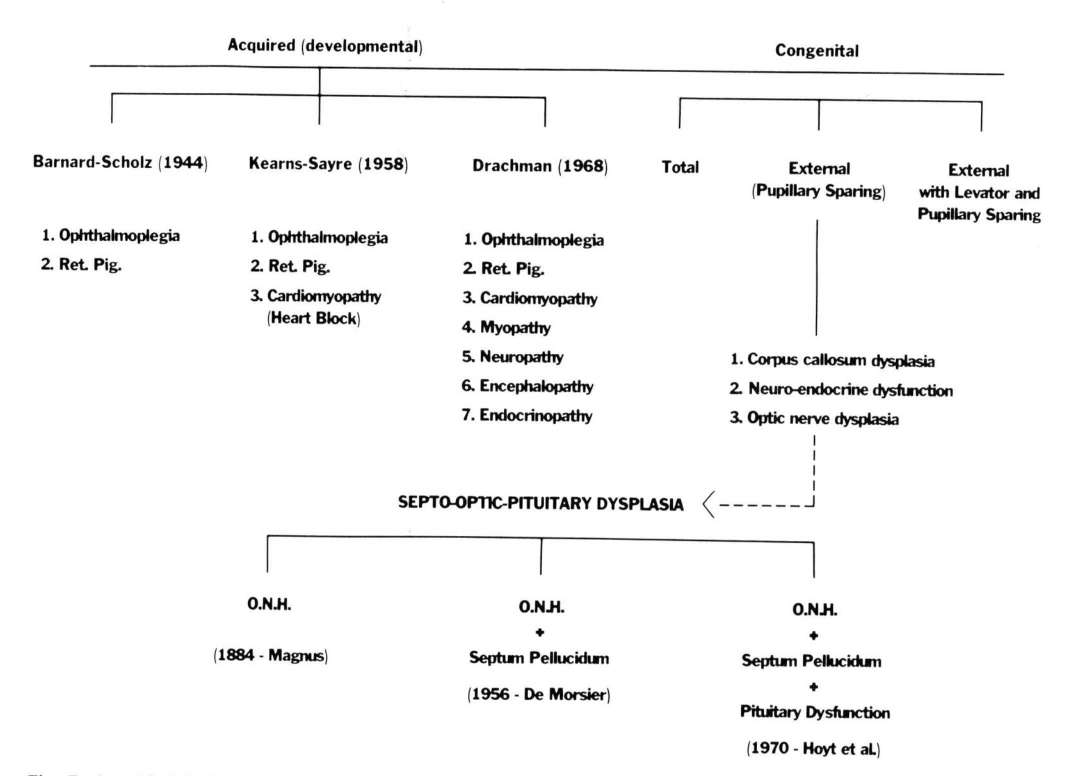

Fig. 7–1. "Ophthalmoplegia plus" syndromes.

nerves, and hypopituitarism has been recognized only recently.

The following embyrogenesis of such a combination of defects is proposed. The cephalad part of the neural tube forms the prosencephalon, mesencephalon, and rhombencephalon. The mesencephalon develops from the walls of the mesencephalic vesicle as the vesicle constricts to form the aqueduct of Sylvius. The tegmentum forms ventral to the aqueduct from the floor of the mesencephalon, and from this area arise the interconnecting nuclear tracts and the nuclei of the oculomotor nerves.

The prosencephalon divides into the proximal diencephalon and the paired vesicles of the distal telencephalon (11-mm stage). The anterior wall of the diencephalon forms the lamina reuniens (16- to 22-mm stage) and from this area, the septum pellucidum and other midline structures later develop (145-mm stage). Their development, however, depends on normal expansion of the cerebral hemispheres and development of the corpus callosum.

In the rostral mesencephalon, the nuclear complex of the oculomotor nerve continues to develop within the gray matter beneath the aqueduct, at the level of the superior colliculi (which perhaps serve as the association center for conjugate vertical eye movements). The nucleus has both paired lateral and unpaired midline subnuclear collections of cells. The levator subnuclear component of the third nerve nuclear complex is an unpaired midline collection of cells located at the caudal end of the nucleus. This cell group may be spared or involved selectively in mesencephalic tegmental lesions. At the cephalic end of the nuclei are the paired dorsomedial Edinger-Westphal nuclei, which provide parasympathetic pupillomotor fibers.

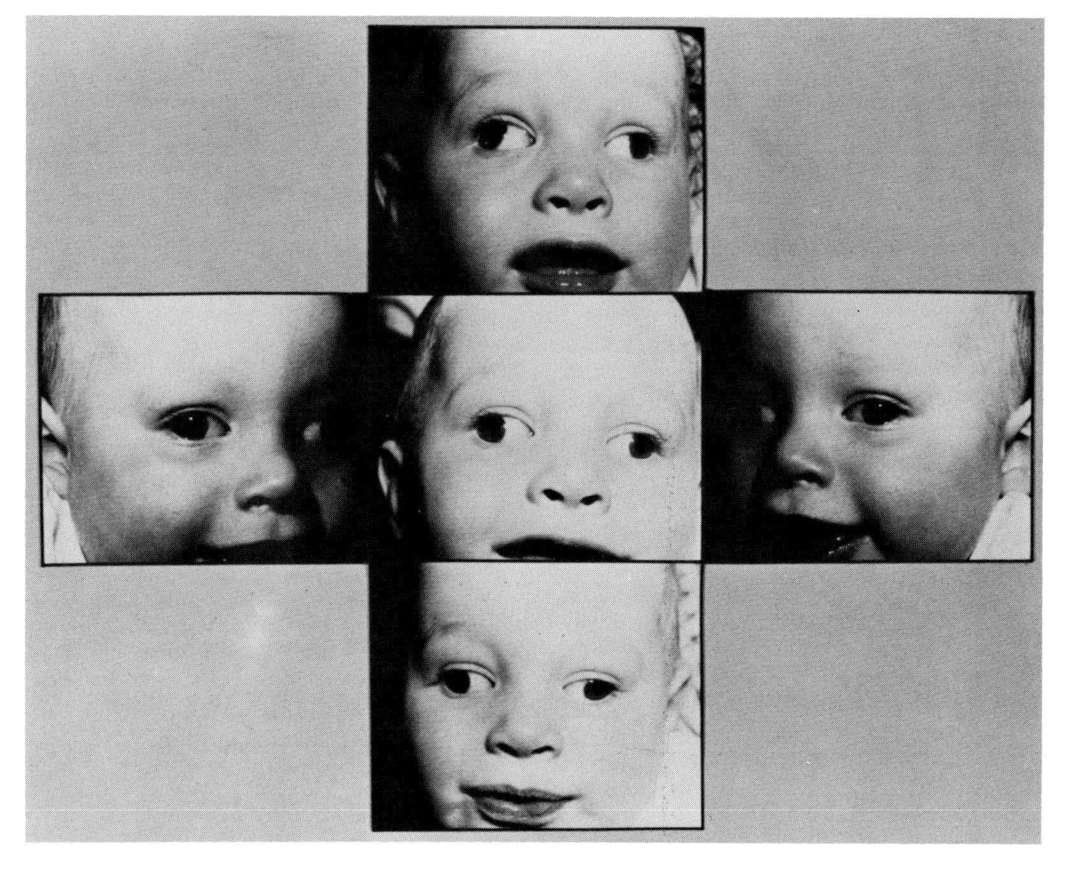

Fig. 7–2. Congenital ophthalmoplegia with lid and pupil sparing.

At some time in the fetal development of the cephalad neural tube—probably between the 11- and 22-mm stages—some type of induction defect may feasibly result in this combination of ophthalmoplegia and dysplasia of the corpus callosum with associated hypopituitarism and optic nerve dysplasia.

Also, this syndrome of congenital ophthalmoplegia/corpus callosum dysplasia/optic nerve dysplasia/pituitary hypofunction (i.e., "congenital ophthalmoplegia plus") may be another form of cephalic dysraphism, akin to the septo-optic-pituitary dysplasia syndrome of Hoyt, Kaplan, and De Morsier, described in Chapter 6.

Of further interest are the increasing descriptions of patients with Duane's syndrome in association with optic nerve dysplasia and neuroendocrine defects. This group of disorders may prove to be another example of combined prosencephalic/mesencephalic dysplasia.

Acers, T.E.: Ophthalmoplegia-corpus callosum dysplasia. Trans. Am. Ophthalmol. Soc., in press.

Acers, T.E.: Optic nerve hypoplasia: The septo-optic-pituitary syndrome. Trans. Am. Ophthalmol. Soc., *79*:425–457, 1981.

Aurand, M.: Ophthalmoplegie externe congenitale bilaterale et familiale. Ann. Ocul., *163*:222, 1926.

Bielschowsky, A.: Congenital oculomotor palsy. *In* Handbuch der Gesamten. 2nd Ed. Leipzig, W. Englemann, 1932, p. 353.

Cincinnati, P., et al.: Holoprosencephaly and agenesis of the corpus callosum. Pediatria (Napoli), *88*:229–247, 1980.

Darsee, J.R., et al.: Mitral valve prolapse and ophthalmoplegia: A progressive cardio-neurologic syndrome. Ann. Int. Med., *92*:735–741, 1980.

DeMorsier, G.: Etudes sur les dystrophies cranioencephaliques. Agenesie du septum lucidum avec malformation du tractus optique: La dysplasia septo-optique. Schweiz. Arch. Neurol. Neurochir. Psychiatr., *77*:267, 1956.

DeSchweinitz, G.E.: Complete bilateral congenital external ophthalmoplegia and ptosis. Trans. Sect. Ophthalmol. AMA, 166–182, 1930.

Goebel, H.H., Komatsuzaki, A., and Bender, M.B.: Lesions of the pontine tegmentum and conjugate gaze paralysis. Arch. Neurol., *24*:431–440, 1971.

Helfand, M.: Congenital familial external ophthalmoplegia without ptosis. Arch. Ophthalmol., *2*:283, 1939.

Holmes, J.W.: Hereditary congenital ophthalmoplegia. Am. J. Ophthalmol., *41*:615–618, 1956.

Hoyt, W.F., et al.: Septo-optic dysplasia and pituitary dwarfism. Lancet, *1*:893–894, 1970.

Jampel, R.S., Okazaki, H., and Bernstein, H.: Ophthalmoplegia and retinal degeneration associated with spinocerebellar ataxia. Arch. Ophthalmol., *66*:247–259, 1961.

Kaplan, S.L., Grumback, M.M., and Hoyt, W.F.: A syndrome of hypopituitary dwarfism, hypoplasia of the optic nerves and malformation of prosencephalon. Pediatr. Res., *4*:480–481, 1970.

Kearns, T.P., and Sayre, G.P.: Retinitis pigmentosa, external ophthalmoplegia and complete heart block. Arch. Ophthalmol., *60*:280–289, 1958.

Li, T.M.: Congenital total bilateral ophthalmoplegia. Am. J. Ophthalmol., *59*:1035–1040, 1965.

McGubbin, S.: Total congenital ophthalmoplegia. Am. J. Ophthalmol., *28*:69, 1941.

Ratzka, M., Orensen, N., and Wodarz, R.: Midline malformations of brain shown by computed tomography. Radiologe, *11*:507–515, 1981.

Reis, R., and Rothfield, S.: Congenital external ophthalmoplegia without ptosis. Albrecht Von Graefes Arch. Klin. Exp. Ophthalmol., *111*:153, 1923.

Salleras, A., and Ortiz De Zarate, J.D.: Recessive sex-linked inheritance of external ophthalmoplegia and myopia, coincident with other dysplasias. Br. J. Ophthalmol., *34*:662–667, 1950.

Schinzel, A.: Postaxial polydactyly, hallux duplication, absence of the corpus callosum, macrencephaly and severe mental retardation: A new syndrome? Helv. Paediatr. Acta, *34*:141–146, 1979.

Slatt, B.: Hereditary external ophthalmoplegia. Am. J. Ophthalmol., *59*:1035–1040, 1965.

Zweifach, P.H., Walton, D.S., and Brown, R.H.: Isolated congenital horizontal gaze paralysis. Arch. Ophthalmol., *81*:345–350, 1969.

Index